In memory of Professor C. D. Collinson

Contents

List of Symbols

A	4-acceleration	$\mathbf{i},\mathbf{j},\mathbf{k}$	unit vectors
A,A_0,x_0	amplitude	i	$\sqrt{-1}$
$\mathbf{a},a,\dot{\mathbf{v}},\ddot{\mathbf{r}}$	acceleration	i	particle index
a	semi-major axis	k	spring constant
$\mathbf{B}$	magnetic field	k_l	linear drag coefficient
b	impact parameter	k_q	quadratic drag coefficient
b	semi-minor axis	$\mathbf{L}$	angular momentum
C	capacitor	$\mathbf{L}^c$	angular momentum vector in centre of mass frame
c	speed of light		
D	directrix	$\mathbf{L}$	total angular momentum of a multi-particle system about the origin
$\mathbf{E}$	electric field		
E	total energy, relativistic energy		
		L	inductor
$\mathbb{E}^3$	Euclidean 3-space	L	Lorentz transformation
$\mathbf{e}_1,\mathbf{e}_2,\mathbf{e}_3$	unit basis vectors	l	semi-latus rectum
e	base of natural logarithms	$\ln$	natural logarithm
e	coefficient of restitution	$\mathbb{M}^4$	Minkowski 4-space
e	eccentricity	M	total mass
e	extension of string	m,m_i	mass
e^-	electron symbol	m_e	rest mass of electron
$\mathbf{F},\mathbf{F}_i,\mathbf{F}_{ij}$	force	m^A	active gravitational mass
$\mathbf{F}_i$	force acting on the ith particle	m^P	passive gravitational mass
		$\mathbf{P}$	total linear momentum
$\mathbf{F}_{ij}$	internal force	$\mathbf{P}^c$	linear momentum vector in centre of mass frame
$\mathbf{F}_i^{\text{ext}}$	external force		
$\mathbf{f}$	frictional force	$\mathbf{P}_i$	linear momentum of ith particle in centre of mass frame
f_T	tension		
$\mathbf{G}$	total external torque	P	4-momentum
G	gravitational constant	$\mathbf{p},\mathbf{p}_i,\mathbf{p}^c$	linear momentum
$g,\mathbf{g}$	gravitational acceleration	$\mathbf{p}$	linear momentum of a particle
h	height		
h	Planck's constant	$\mathbf{p}$	relativistic 3-momentum
$\hbar$	$h \div 2\pi$	$\bar{p}$	antiproton symbol
I	3×3 identity matrix	Q,q	charge
I	alternating current	$q,q^c,\mathbf{q}$	linear momentum
I,I^c	moments of inertia	$\mathbf{R}$	3×3 rotation matrix

$\mathbf{R}^{\mathrm{T}}$	3×3 transposed matrix	ϵ	permittivity
$\mathbf{R}$	position of centre of mass	ϵ	small displacement
$\mathbf{R}$	resultant reaction	$\zeta(t)$	complex variable
R	radius of Earth	η	constant viscosity coefficient
R	horizontal range	η	Minkowski metric
R	resistor	$\Theta, \theta, \theta^{\mathrm{c}}$	scattering angle
R_{e}	Reynolds number	θ	polar angle
$\mathbf{r}, \mathbf{r}_i$	position vector	$\hat{\boldsymbol{\theta}}$	transverse unit vector
$\mathbf{r}_i^{\mathrm{c}}$	position vector in centre of mass frame	$\dot{\theta}, \omega$	angular speed
$\hat{\mathbf{r}}$	radial unit vector	λ	latitude
$\mathbf{r}, \mathbf{r}_{ij}$	relative position	λ	wavelength
r	radial coordinate	μ_{k}	coefficient of kinetic friction
S, S'	inertial frame	μ_{s}	coefficient of static friction
T	rotational kinetic energy	μ	permeability
T, T^{c}	kinetic energy	μ	reduced mass
t	time	ν	frequency
U	4-velocity	ν_0	proper frequency
$\mathbf{u}$	3-velocity	ρ	density
$\mathbf{v}, \mathbf{v}_i, \dot{\mathbf{r}}$	velocity	ρ	rapidity
V	4-vector	τ_{p}	period of small oscillations
V	potential energy	τ	periodic time
V	volume	τ	proper time
V_{alt}	voltage	τ	time of flight
V_{eff}	effective potential energy	ϕ, ϕ^{c}	recoil angle
V^{int}	internal potential energy	ξ	small perturbation
V^{ext}	external potential energy	ω	angular frequency
v_∞	terminal speed	ω	cyclotron frequency
W	work	$\omega, \boldsymbol{\Omega}$	angular velocity
X	reactance	ω_0	resonant frequency
X	position 4-vector	$\mathbf{u} \cdot \mathbf{v}$	inner (scalar) product
$\mathbf{x}$	position 3-vector	$\mathbf{u} \times \mathbf{v}$	vector cross product
x, y, z	Cartesian coordinates	$\mathbf{u} \uparrow\uparrow \mathbf{v}$	parallel vectors
Z	impedance	∇	vector differential operator
γ	Lorentz factor	$*$	complex conjugation
γ	photon symbol	$\square$	end of example

Preface

This book is intended as an introduction to the essential aspects of dynamics and special relativity. It has been designed to be readily accessible to students studying both of these subjects for the first time, and incorporates only those elements that would be considered as essential for any undergraduate course in which these subjects are featured. The examples and exercises are an integral part of the book and the reader should spend a little time working through them.

My main motivation for writing this book comes from the desire to address a number of issues that have plagued readers over the years. From a mathematical perspective, dynamics is not difficult. Indeed, very little mathematics beyond the standard calculus with some basic knowledge of vectors and matrices is assumed or required. From a physical perspective, dynamics is not difficult. The reason I say this is because unlike subjects such as quantum mechanics and, to some extent, special relativity, that deal with entities that are difficult to comprehend, dynamics, as presented here, will concentrate on macroscopic quantities that travel at speeds well below that of light. In this sense, dynamics is a very intuitive and readily amenable subject. Why, then, do students who first encounter dynamics usually find the subject challenging? The main reason is due to the vast array of new physical concepts that any novice will encounter. Each physical concept comes with a very precise definition that must be strictly adhered to for the physics to bear out the mathematics. In everyday life we use dynamical terms very loosely and without much thought. However, when studying dynamics for the first time, we must force ourselves out of the habit of discussing *weight* when we really mean *mass*, or *velocity* when we really mean *speed*, and so the list goes on. Once we begin to think in precise terms and to use precise nomenclature, the difficulty associated with comprehending the physics will vanish. In an attempt to resolve this problem for the reader, I have been careful to repeat certain important ideas and terms throughout the book as well as filling in gaps that tend to be considered as obvious, and are, therefore, often omitted. Another cause of confusion can arise from the way in which dynamics is presented. Often a topic such as *energy* will be discussed by bringing together other areas of dynamics, such as *rigid bodies, simple harmonic motion*, etc., to illustrate the concept. This can be a good thing to do *if* the reader is already aware of the technicalities involved here. However, I have decided to present the subject whereby the reader first meets a dynamical topic, *rigid bodies* say, and then the ideas of *energy, angular momentum*, etc. are developed within that topic. This should comfortably place each dynamical topic within the wider constructs of the theory, without the necessity to jump from topic to topic to illustrate a point.

In the past, people found it very difficult to decide where best to place special relativity. It would be subsumed within courses on electromagnetism, general relativity and, sometimes, quantum mechanics. It is, of course, relevant to all those subjects, but it is my view (and many other lecturers of the subject will agree) that special relativity is best juxtaposed with dynamics. Indeed, as we will see, classical dynamics is recovered from special relativity for velocities much less than the speed of light. Unlike dynamics, special relativity is not intuitive — especially for someone new to the subject. Confusion arises because we can no longer visualise the various processes within this theory — one must, invariably, rely on what the mathematics is telling us, although extremely ingenious experiments have been conducted (and are being conducted) to verify all of the non-intuitive ideas that form special relativity. Nevertheless, the mathematics is again straightforward and as we shall see there are some visual aids that will prove useful if used with care. Again, we will consider only those aspects of the theory that are considered to be essential elements. Once understood, readers will view the world quite differently!

The book is split into two sections: Dynamics and Relativity. The first three chapters concentrate on the fundamental aspects of single-particle motion. The next three chapters begin to analyse the motion of more than a single particle, culminating in Chapter 6 when multi-particle systems are discussed. Chapter 7 uses the ideas of Chapter 6 to enable one to understand the dynamical behaviour of composite bodies (*rigid bodies*). In Chapter 8, Newton's laws will have to be reformulated for them to work in non-inertial frames of reference (non-accelerating frames of reference). This chapter analyses the dynamical behaviour of objects in non-inertial frames of reference; specifically, *rotating reference frames*. The section on relativity comprises three chapters. Chapter 9 discusses the ideas leading to the inception of special relativity and formulates those fundamental aspects, such as *time dilation, length contraction*, the *Lorentz transformations* and the visual aids of *Minkowski diagrams*, necessary to develop more sophisticated ideas in the following chapters. Chapter 10 introduces a very important quantity called the *4-vector*, and Chapter 11 presents the ideas thus far developed within the context of relativistic mechanics.

Peter J. O'Donnell
Cambridge
U.K.

About the Author

Peter O'Donnell lectures in the Department of Applied Mathematics and Theoretical Physics, University of Cambridge, and is a member of its Relativity and Gravitation Research Group. His research interests are in general relativity, twistor theory and quantum gravity. He is the author of a number of research papers in these fields, as well as being the author of the critically acclaimed monograph 'An Introduction to 2-spinors in General Relativity' (World Scientific, 2003). He is also a Fellow and Director of Studies for Mathematics at St. Edmund's College, Cambridge. He has lectured courses on both classical mechanics and special relativity over many years.

Acknowledgments

I would like to thank the following people for looking at various chapters of the book and for their helpful comments and suggestions: Professor John Barrow, Dr. James Foster and Professor David Tong. I am most indebted to Professor Graham Hall and Dr. Clive Wells, who painstakingly reviewed each chapter. I would also like to thank Dr. Jennifer Atkinson for additional proofreading and Francesca McGowan for her editorial assistance.

1

The Galileo–Newton Formulation of
Dynamics

CONTENTS

This chapter is primarily concerned with establishing the crucial ideas and concepts that form the bedrock of what has come to be known as dynamics. Primarily, but not exclusively, from the adroit minds of Galileo and Newton, beautifully crafted notions of the way the universe functions were laid down with precise mathematical elegance, for all to witness. However, the simplicity and clarity of these expositions were not always accepted by either of these two great scientists' contemporaries. Indeed, Galileo especially was victimised for many of his scientific views, which were later to form the basis of Newton's development of dynamics. It will prove instructive if we first take a glimpse at how the ancients perceived their universe.

The Aristotlean theory of the universe may appear fantastic and naive from a 21st century perspective; nevertheless, the ancient Greeks formulated a superb understanding of *statics* — now a branch of classical mechanics that deals with bodies in dynamical equilibrium. However, their cognisance of moving bodies was extremely limited. Yet even Newton was unwilling to abandon all Aristotlean notions when it came to formulating his own laws of motion.

Aristotle had a firm belief that the universe was composed of five basic elements: earth, water, air, fire and aether. The first four elements of his physical model were arranged as concentric spheres with the Earth being the central most sphere. Aether formed the *celestial sphere* in which resided the Moon, stars and all other *heavenly bodies*. Aristotle's *geocentric* model of the universe attempted to elucidate upon a number of burning issues that occupied classical Greek thought; not least, why bodies made from the element earth are necessarily attracted to the Earth, which could be considered as the first step towards an explanation of gravitational attraction of massive bodies. The important aspect to all this from our perspective is the idea that material bodies would tend towards the centre of this geocentric universe. In this context, space has a preferred origin: *space is absolute*. In other words, only at this preferred spatial origin is a material body considered to be truly at rest.

1.1   Galilean relativity

The concept of absolute space continued through the 17th century and into the very early part of the 20th century. Indeed, Newton was reluctant to dismiss its existence whilst formulating his dynamical laws, requiring a quasi-absolute space (centre of mass of the solar system) with respect to which the dynamical behaviour of bodies can be described.

Galileo had realised in 1638 that one's position in space is merely a relative concept. His reasoning went along the following lines. Consider yourself aboard a ship, tucked away inside a cabin with no view of the outside world. Then if the sea was calm it would be impossible, in principle, to ascertain whether the ship was stationary or moving with constant speed. This statement embodies the *Galilean principle of relativity*. Thus there is no dynamically preferred state of rest.

1.1.1   Frames of reference

The framework of classical dynamics as devised by Galileo and Newton consists of *particles* moving in a three-dimensional space governed by Euclidean geometry. In this space, which we shall refer to as Euclidean 3-space $\mathbb{E}^3$, a particle's dynamical behaviour is determined by the way in which its position changes with time. A *particle* is defined as a point-like mass: a spatial point with mass. It possesses no other internal physical attributes and is impervious to deformations. The only distinguishable property a particle has is its *mass* — this notion will be discussed in Section 1.3. Particles can be combined together to form *a system of particles* or *extended bodies* such as laminae, rocks and planets. Indeed, we will observe in Chapter 6 that an extended body can be treated, to a very good approximation, as a single particle; this is provided that one is merely interested in the path of the body through space and not in other aspects of its dynamical behaviour; for example, its rotational motion.

It is important to recognise that as particles move through space they are also moving through time. In the Galileo–Newton formulation of classical dynamics, both space and time are conventionally treated as separate entities within the fabric of the physical universe. (We will present a slightly modified version when discussing special relativity in Chapter 9. There, space and time are combined into an all encompassing *space-time*.)

Consider a particle travelling along a path C in Euclidean 3-space $\mathbb{E}^3$ (Figure 1.1). The point P is the position of the particle relative to an origin O at some time t. The vector $\mathbf{r}$ is the *position vector* of P relative to O. The choice of coordinates that can be employed to specify $\mathbf{r}$ is by no means unique. However, it will suit our purpose, for the present, to choose a *right-handed* (rectangular) *Cartesian coordinate system* as a *frame of reference* or *frame*. Thus the position of the particle at P is specified by coordinates (x, y, z) relative to a given set of mutually orthogonal axes Ox, Oy and Oz. The *basis vectors* associated with our Cartesian coordinate system are denoted by $\mathbf{i}, \mathbf{j}$ and $\mathbf{k}$ and represent the *unit vectors* parallel to the axes Ox, Oy and Oz, respectively. Accordingly, one can write the position vector $\mathbf{r}$ in terms of Cartesian components (x, y, z) as

$$\mathbf{r} = x\mathbf{i} + y\mathbf{j} + z\mathbf{k}.$$

The notation $\mathbf{r} = (x, y, z)$ will be used frequently as a convenient abbreviation for the *components* of $\mathbf{r}$ in those circumstances where there is no possibility of ambiguity.

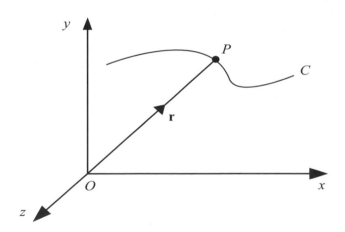

FIGURE 1.1
Particle along a path

Now as time varies, the particle's *trajectory* is merely the equation of the path C parameterised, along C, by t:

$$\mathbf{r} = \mathbf{r}(t).$$

One can think of oneself as an *observer* positioned at the origin O of the frame in Figure 1.1. As the particle traverses the path C the position vector $\mathbf{r}$ will change with time. Thus, the *velocity* $\mathbf{v}$ of the particle is defined as *the rate of change of* $\mathbf{r}$ *with respect to* t:

$$\mathbf{v} = \frac{d\mathbf{r}}{dt} \equiv \dot{\mathbf{r}}.$$

(The dot above the letter will be used as a convenient abbreviation to denote the derivative of a quantity with respect to *time*.) The velocity $\mathbf{v}$ is defined to be a *constant* or *uniform velocity* if and only if its *speed* $|\mathbf{v}|$ (magnitude of $\mathbf{v}$) and *direction* are both constant. This means that the locus of the particle's trajectory will be a straight line relative to the observer at O, or a point if $\mathbf{v} = \mathbf{0}$.

The *acceleration* $\mathbf{a}$ of the particle is defined as *the rate of change of* $\mathbf{v}$ *with respect to* t:

$$\mathbf{a} = \frac{d\mathbf{v}}{dt} \equiv \dot{\mathbf{v}}$$
$$= \frac{d^2\mathbf{r}}{dt^2} \equiv \ddot{\mathbf{r}}.$$

It is clear that for a particle moving with uniform velocity ($\mathbf{v}$ is constant relative to the observer at O) the acceleration of the particle will be zero. The converse of this statement is also true.

Our discussion thus far has been based on the fact that an *event* (some point in space and time) has occurred relative to an origin O. Namely, an observer at O witnesses the motion of a particle. Now consider a second observer positioned at a different origin, O', say. How does the velocity and acceleration of the particle differ for each observer?

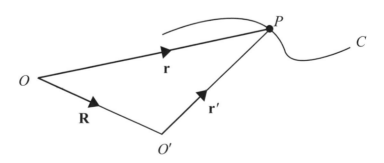

FIGURE 1.2
Relative position of origins

Let $\mathbf{r}'$ be the position vector of P relative to O' and $\mathbf{R}$ the position vector of O' relative to O. Thence from Figure 1.2

$$\mathbf{r}' = \mathbf{r} - \mathbf{R}. \tag{1.1}$$

If we assume that both observers synchronise their clocks such that $t' = t$, then the velocity of P relative to O' is

$$\mathbf{v}' = \frac{d\mathbf{r}'}{dt'} = \frac{d\mathbf{r}'}{dt}.$$

Thus differentiating (1.1) with respect to t yields

$$\mathbf{v}' = \mathbf{v} - \mathbf{V}, \tag{1.2}$$

where $\mathbf{V} = d\mathbf{R}/dt$ is the velocity of the observer at O' relative to the observer at O. Notice that if $\mathbf{R}$ is a constant vector then $\mathbf{V} = \mathbf{0}$ and the velocity of the particle is independent of origin. The relation for the acceleration of the particle relative to O and O' is obtained by differentiating (1.2) with respect to time:

$$\dot{\mathbf{v}}' = \dot{\mathbf{v}} - \dot{\mathbf{V}}. \tag{1.3}$$

Had the clocks of our two observers been synchronised — that is, if the observer at O' started his clock t_0 seconds after the observer at O started his clock — then (1.2) and (1.3) would be unaffected. In other words, (1.2) and (1.3) remain *invariant* under time translations of the form

$$t' = t + t_0.$$

Thus regardless of whether O' is stationary or moving with uniform velocity relative to O, both observers will be in total agreement as to the exact time at which an event takes place: *time is absolute*.

It should be pointed out that when we come to discuss special relativity the notions of absolute space and absolute time must necessarily be abandoned. In that theory, space and time are merely *relative* quantities.

Example 1.1 A point P travels through an origin O at time $t = 0$. Its rectilinear (straight-line) motion is governed by the equation

$$\ddot{x}(t) - k\dot{x}(t) = 0,$$

where k is a non-zero constant with units s^{-1}. Given initial conditions $\dot{x}(0) = v_0 > 0$ and $x(0) = 0$, find the velocity and position of P at later t.

Solution The given equation can be written as

$$\frac{d\dot{x}}{dt} = k\dot{x}.$$

This is a first-order separable equation in $\dot{x}$. Integrating yields

$$\int \frac{d\dot{x}}{\dot{x}} = \int k\,dt + \ln|C_1|,$$

where $\ln|C_1|$ has been chosen for simplicity as the constant of integration. Thus

$$\dot{x} = C_1 e^{kt}.$$

The first initial condition gives $C_1 = v_0$. Therefore,

$$\dot{x}(t) = v_0 e^{kt}.$$

Integrating this gives

$$x(t) = \frac{v_0}{k} e^{kt} + C_2.$$

The second initial condition gives $C_2 = -v_0/k$. Therefore,

$$x(t) = \frac{v_0}{k}(e^{kt} - 1). \qquad \square$$

Example 1.2 A point P can travel along the x-axis such that its position (in metres) at any given time t is

$$x(t) = t^3 - 9t^2 + 15t + 18.$$

Find the instantaneous velocity and acceleration of P at $t = 2$ s and comment on the result. Give the times at which P is at rest.

Somewhere along its path P twice changes direction. Comment on P's direction, velocity, acceleration and speed over the interval $-\infty < t < \infty$.

Solution The velocity and acceleration of P are, respectively,

$$\dot{x}(t) = 3(t - 1)(t - 5)$$

and

$$\ddot{x}(t) = 6(t - 3).$$

At $t = 2$, $\dot{x}(2) = -9$ and $\ddot{x}(2) = -6$. So P is moving *left* (negatively) along the x-axis at a *speed* of 9 ms^{-1}. The instantaneous velocity is, therefore, -9 ms^{-1}. However, as the acceleration is also negative, P must be increasing in speed. When P is at rest the velocity must be zero. This occurs at $t = 1$ and $t = 5$. When the velocity of P is constant the acceleration must be zero. This occurs at $t = 3$.

Now, the graph of $x(t)$ against t has two turning points. These are obtained by putting $\dot{x}(t) = 0$. We can use the acceleration equation to determine the nature of the turning points. Thus

at $t = 1$, $\ddot{x}(1) = -12$ $\therefore$ max. turning point at $t = 5$, $\ddot{x}(5) = +12$ $\therefore$ min. turning point.

Also, $\ddot{x}(t) < 0$ for $t < 0$ and $\ddot{x}(t) > 0$ for $t > 0$. We will gain a better understanding of the dynamical behaviour of P by combining all of the above information in tabular form (see Table 1.1).

TABLE 1.1
Dynamical behaviour of P

$-\infty < t < 1$	$\dot{x}(t) > 0$	$\ddot{x}(t) < 0$	$\lvert\dot{x}(t)\rvert$ decreasing
$1 < t < 3$	$\dot{x}(t) < 0$	$\ddot{x}(t) < 0$	$\lvert\dot{x}(t)\rvert$ increasing
$3 < t < 5$	$\dot{x}(t) < 0$	$\ddot{x}(t) > 0$	$\lvert\dot{x}(t)\rvert$ decreasing
$5 < t < \infty$	$\dot{x}(t) > 0$	$\ddot{x}(t) > 0$	$\lvert\dot{x}(t)\rvert$ increasing

As P moves *right* (positively) along the x-axis it experiences a change of direction just after $t = 1$ and again just after $t = 5$. Notice that if the velocity and acceleration are opposite in sign the speed of P will decrease. Also, if the velocity is zero it is not necessarily true that the acceleration will be zero (this is discussed in more detail in Section 2.2). □

Example 1.3 A particle is projected vertically upwards. During its *time of flight* it reaches a height

$$y(t) = -\frac{1}{2}gt^2 + C_1 t + C_2$$

above the ground, where g is the constant *acceleration due to gravity* ($g \approx 9.8$ ms^{-2}), and C_1 and C_2 are arbitrary constants. For the given conditions $y(t_0) = h$ and $\dot{y}(t_0) = v_0$, find the velocity of the particle as it hits the ground.

Solution The velocity of the particle is

$$\dot{y}(t) = -gt + C_1.$$

Substituting the given conditions into y and $\dot{y}$, respectively, yields

$$h = -\frac{1}{2}gt_0^2 + C_1 t_0 + C_2$$

and

$$v_0 = -gt_0 + C_1,$$

so C_1 and C_2 can now be found explicitly. Thus the y and $\dot{y}$ equations are

$$y(t) = h - \frac{1}{2}g(t - t_0)^2 + v_0(t - t_0)$$

and

$$\dot{y}(t) = v_0 - g(t - t_0).$$

The particle hits the ground at $y = 0$. This occurs at some time $t = t_1$, say. Thus, we are required to solve the quadratic

$$\frac{1}{2}g(t_1 - t_0)^2 - v_0(t_1 - t_0) - h = 0.$$

The roots of this equation are

$$t_1 - t_0 = \left(v_0 \pm \sqrt{v_0^2 + 2gh}\right)\Big/ g.$$

As t_1 is a later time than t_0, $t_1 - t_0 > 0$. Hence,

$$\dot{y}(t_1) = v_0 - \left(v_0 + \sqrt{v_0^2 + 2gh}\right)$$
$$= -\sqrt{v_0^2 + 2gh},$$

which is the velocity of the particle as it hits the ground. $\qquad\square$

Example 1.4 A passenger on a train is travelling at 100 kmh^{-1} in a northeasterly direction. He watches an aeroplane that appears to be flying at 200 kmh^{-1} due west. Find the speed relative to the ground and direction (*bearing*) of the aeroplane.

Solution We can form a vector triangle (see Figure 1.3) consisting of the velocity of the train $\mathbf{v_T}$, the actual velocity of the aeroplane $\mathbf{v_A}$, and the velocity $\mathbf{v_{A-T}}$ of the aeroplane relative to the train: $\mathbf{v_{A-T}} = \mathbf{v_A} - \mathbf{v_T}$. Then the cosine rule can be used to determine the speed of the aeroplane relative to the ground. Hence,

$$|\mathbf{v_A}|^2 = |\mathbf{v_{A-T}}|^2 + |\mathbf{v_T}|^2 - 2|\mathbf{v_{A-T}}||\mathbf{v_T}|\cos(45°)$$
$$= (200)^2 + (100)^2 - 2(200)(100)(1/\sqrt{2})$$
$$= 40000 + 10000 - 40000/\sqrt{2}$$
$$= 21715.7$$
$$\therefore \ |\mathbf{v_A}| = 147.4 \text{ kmh}^{-1}.$$

The sine rule can be used to determine the actual bearing of the aeroplane. Hence

$$\theta = \sin^{-1}\left(\frac{100/\sqrt{2}}{147.4}\right) = 28.7°$$

implying that its bearing is $270° + 28.7° = 298.7°$. $\qquad\square$

1.1.2 Galileo's law of inertia and inertial frames

Recall the Galilean principle of relativity in Section 1.1. What was not mentioned there was that eventually the ship, if moving relative to the water,

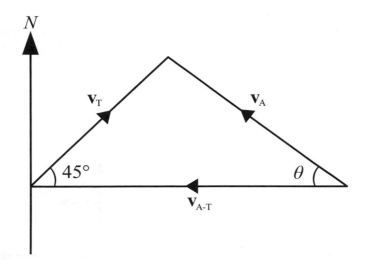

FIGURE 1.3
$v_{A-T} = v_A - v_T$

would come to rest in the water. Why should this be? Aristotle would argue that it is because the natural state of any body is that of rest. However, Galileo brilliantly recognised that: *any particle freely moving* (under the influence of no force) *would naturally continue in a straight line with constant speed.* This is *Galileo's law of inertia* postulated in 1638. Of course, the ship *is* under the influence of a force: *friction.* The frictional force here arises from the flow of water and air around the ship.

Any frame of reference in which Galileo's law of inertia holds is referred to as an *inertial frame.* These frames are extremely important in dynamics because Newton's dynamical laws hold only in inertial frames without need of modification. Frames in which the law of inertia does not hold — for example, rotating or linearly accelerating frames — are known as *non-inertial*; Newton's dynamical laws must be rephrased in these frames.

It should be borne in mind that the concept of an inertial frame is purely idealistic. For a body to experience no force, and therefore to be moving freely, one would require the body to be completely isolated. In practice this requirement is not possible. Nevertheless, we can conceive of frames that are very close to becoming inertial.

To a very good approximation one can consider a frame fixed to the surface of the Earth as inertial relative to the distant stars, which are themselves considered as fixed.[1] This means that if a dynamical experiment is conducted

---

[1]Not only are the distant stars in motion but the Earth, of course, rotates about its axis

by an observer in an inertial frame S then observers in an inertial frame S', moving freely with respect to S, will agree with any conclusions reached by the observer in S. In other words, the laws of physics remain invariant with respect to any inertial frame. The pronoun *any* is important here. Inertial frames are not unique — there are infinitely many.

1.1.3   Galilean transformations

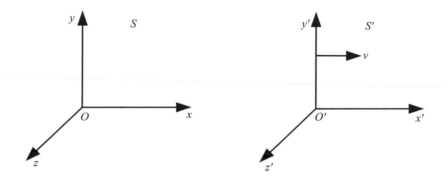

FIGURE 1.4
Two frames moving relative to one another

It has already been stated that the laws of physics are the same in any inertial frame. Thus, it would be desirable to relate observations made in one inertial frame to that of another inertial frame. Consider two such frames, S with coordinates (x, y, z, t) and S' with coordinates (x', y', z', t') as depicted in Figure 1.4. We will adopt the *standard configuration* to orientate the frames as follows. An observer at O' in the S' frame moves at constant speed v relative to an observer at O in the S frame and along the x-axis. Both observers synchronise their clocks so that at $t = t' = 0$ both frames coincide: $x = x'$, $y = y'$ and $z = z'$.

Now, imagine an event such as a torch flashing. The observer in S' will witness the event at the point (x', y', z', t') while the observer in S will witness the event at the point (x, y, z, t) (see Figure 1.5). However, as time is considered to be absolute then observers O' and O will both record the time of the event at $t' = t$. As motion is confined along the x-axis, then at subsequent times, $y' = y$ and $z' = z$. Of course, $x' \neq x$.

The flashing torch, which is *fixed* in S', is not fixed in S, so at the moment

---

while orbiting the Sun. The non-inertial character of 'Earth-frames' is manifest in the area of ballistics, e.g., high velocity projectiles deviate from their straight-line paths over large distances.

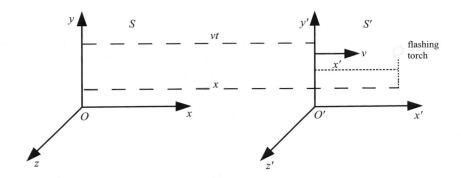

FIGURE 1.5
Two frames moving relative to one another with an event

the torch flashes the observers will be separated by a distance $x' + vt$. Finally, the relationship between coordinate systems in S' and S is given by

$$
\begin{aligned}
x' &= x - vt \\
y' &= y \\
z' &= z \\
t' &= t.
\end{aligned}
\tag{1.4}
$$

These are the *Galilean transformations*. One can represent the Galilean transformations in vector notation, which is sometimes more convenient. Thus, defining $\mathbf{r}' = (x', y', z')$ in S' and $\mathbf{r} = (x, y, z)$ in S, and allowing $t = t'$ to act as a parameter, (1.4) can be written as

$$
\mathbf{r}' = \mathbf{r} - \mathbf{v}t,
\tag{1.5}
$$

where $\mathbf{v} = (v, 0, 0)$. This is commonly referred to as a *Galilean boost*.

Next, consider the torch moving at uniform velocity $\mathbf{u}' = (\dot{x}', \dot{y}', \dot{z}')$ in S' and $\mathbf{u} = (\dot{x}, \dot{y}, \dot{z})$ in S. Differentiating (1.5) with respect to time yields

$$
\dot{\mathbf{r}}' = \dot{\mathbf{r}} - \mathbf{v}
$$

or

$$
\mathbf{u}' = \mathbf{u} - \mathbf{v},
\tag{1.6}
$$

which is merely the relationship for the relative velocity of our two observers exhibited in (1.2). Bearing in mind that the relative velocity is uniform, and therefore the vector $\mathbf{v}$ is constant, a second differentiation with respect to time yields

$$
\dot{\mathbf{u}}' = \dot{\mathbf{u}}
$$

or

$$\mathbf{a'} = \mathbf{a}.$$

Therefore, both *inertial observers* will yield the same result on measuring the acceleration of the flashing torch in their respective inertial frames. *Acceleration is invariant under Galilean transformations.*

> **Example 1.5** Burger's equation[2] is a non-linear partial differential equation used extensively in *fluid dynamics* and is given in one dimension by
>
> $$\frac{\partial u}{\partial t} + u\frac{\partial u}{\partial x} = \eta\frac{\partial^2 u}{\partial x^2},$$
>
> where $u = u(x,t)$ is the one-dimensional velocity and η is the constant viscosity coefficient. Show that this equation remains invariant under Galilean transformations.
>
> **Solution** We wish to transform Burger's equation from the unprimed frame to the primed frame and compare the two equations.
>
> On employing the chain rule for a multivariate function, we obtain relations for $\frac{\partial u}{\partial x}$, $\frac{\partial^2 u}{\partial x^2}$ and $\frac{\partial u}{\partial t}$:
>
> $$\begin{aligned}
\frac{\partial u}{\partial x} &= \frac{\partial u}{\partial x'}\frac{\partial x'}{\partial x} + \frac{\partial u}{\partial t'}\frac{\partial t'}{\partial x} \\
&= \frac{\partial u}{\partial x'} + 0 \quad \text{from (1.4)} \\
&= \frac{\partial}{\partial x'}(u' + v) \quad \text{from (1.6)} \\
&= \frac{\partial u'}{\partial x'}
\end{aligned}$$
>
> (as v is the constant velocity of one frame relative to the other),
>
> $$\frac{\partial^2 u}{\partial x^2} = \frac{\partial^2 u'}{\partial x'^2}$$
>
> and
>
> $$\begin{aligned}
\frac{\partial u}{\partial t} &= \frac{\partial u}{\partial x'}\frac{\partial x'}{\partial t} + \frac{\partial u}{\partial t'}\frac{\partial t'}{\partial t} \\
&= -v\frac{\partial u}{\partial x'} + \frac{\partial u}{\partial t'} \quad \text{from (1.4)} \\
&= -v\frac{\partial}{\partial x'}(u' + v) + \frac{\partial}{\partial t'}(u' + v) \quad \text{from (1.6)} \\
&= -v\frac{\partial u'}{\partial x'} + \frac{\partial u'}{\partial t'}.
\end{aligned}$$

---

[2] Johannes Martinius Burger (1895-1981)

Substituting the above relations into Burger's equation yields

$$-v\frac{\partial u'}{\partial x'} + \frac{\partial u'}{\partial t'} + (u' + v)\frac{\partial u'}{\partial x'} = \eta\frac{\partial^2 u'}{\partial x'^2}$$

or

$$\frac{\partial u'}{\partial t'} + u'\frac{\partial u'}{\partial x'} = \eta\frac{\partial^2 u'}{\partial x'^2}$$

as required. □

1.1.4 General Galilean transformations

It is worth mentioning that the Galilean transformations given by (1.4) are not the most general type that would allow one to transform between inertial frames. If transformations such as coordinate and time rescalings are not considered,[3] there are three transformations which in combination would constitute the most general transformation.

For the sake of brevity, we will present these transformations in the form of a general coordinate $(\mathbf{r}, t)$ in 'Galilean space-time' ($\mathbb{E}^3$ in combination with a time coordinate). Also, these transformations are by convention formulated for a fixed observer in S' relative to a moving observer in S. This will account for the visible difference in sign in what follows.

We have already encountered the boost transformation. The name arises due to the fact that an inertial frame is given a 'boost' in a particular direction; in this case it is in the x-direction. This is also the only transformation in our set of four that exhibits the time t explicitly in the space part of the space-time coordinates:

$$(\mathbf{r}', t') = (\mathbf{r} + \mathbf{v}t, t).$$

The fact that the space $\mathbb{E}^3$ is *homogeneous* (at any moment in time the universe appears the same at any point in space) is described by the combined spatial and temporal translations:

$$(\mathbf{r}', t') = (\mathbf{r} + \mathbf{r}_0, t + t_0),$$

where $\mathbf{r}_0$ and t_0 are constants. $\mathbb{E}^3$ is also *isotropic* (the universe appears the same in any direction) as described by the *rotations*:

$$(\mathbf{r}', t') = (R\mathbf{r}, t), \tag{1.7}$$

where R is a 3×3 rotation matrix (orthogonal matrix with $\det R = 1$). Then the axes of S could be rotated by an angle θ so that they are no longer parallel to the axes of S' (see Figure 1.6). For example, if a rotation is performed

---

[3]This type of transformation merely allows one to discuss a change of units between inertial frames; for example, kilometres in S' and miles in S.

about the z-axis by an angle θ in the anticlockwise direction, the matrix $R(\theta)$ representing this is

$$R(\theta) = \begin{pmatrix} \cos(\theta) & -\sin(\theta) & 0 \\ \sin(\theta) & \cos(\theta) & 0 \\ 0 & 0 & 1 \end{pmatrix}.$$

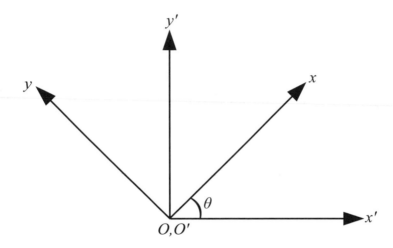

FIGURE 1.6
Rotating frames

    Note that the same transformation (1.7) would allow for *reflections* (det $R = -1$). However, the transformation would result in going from a right- to a left-handed coordinate system, which is a discrete rather than continuous process.

    Thence, the most general transformation between inertial frames is given by a combination of the above three transformations:

$$(\mathbf{r}', t') = (R\mathbf{r} + \mathbf{v}t + \mathbf{r}_0, t + t_0).$$

1.2   Newton's dynamical laws

In his great work of 1687, the *Philosophiae Naturalis Principia Mathematica*, Sir Isaac Newton (1642-1727) propounded three axioms that are arguably the three most cardinal laws of the nature of the universe. They have come to be known as *Newton's three laws of motion*[4]:

(NI) *Newton's first law (Galileo's law of inertia).* Every particle continues in a state of rest, or of uniform motion in a straight line, unless acted upon by force.

(NII) *Newton's second law.* The acceleration (or rate of change of momentum) of a particle is proportional to the force acting on the particle.

(NIII) *Newton's third law.* For every action, there is always an equal and opposite reaction.

Like Galileo, and Aristotle before him, Newton also introduced the concepts of absolute time and absolute space to his mechanical model of the universe. That time was a quantity that could be measured accurately and without ambiguity given observers with good-working clocks was taken for granted by Newton and his contemporaries. However, Newton's claim that space too should be considered as absolute had to be more forcibly made to convince a small but elite band of adversaries. His hypothesis was based on the assumption that a preferred state of absolute rest with regard to linear motion exists and that its location was at the *centre of gravity* of the solar system. He conceded the point, however, that determining such a state of rest by observation would prove problematic. Nevertheless, Newton's dynamical laws do not justify the concept of absolute space nor are they inconsistent with it.

Newton's first law of motion arises qualitatively from profound thinking. It is nevertheless a claim that would be difficult to prove in practice, but the validity of which has stood the test of time and is unlikely to be refuted. On the other hand, Newton's second law of motion arises empirically through experimentation. On comparing these two laws, one might suggest that the first law is merely a special case of the second law. That is, in the absence of force the acceleration of a particle vanishes and that particle will continue at a uniform velocity, or remain at rest; which is just a statement of Newton's first law! Essentially, the first law is telling us: *inertial frames exist.* The second law can then be experimentally verified in these inertial frames and the concept of force can then be defined in terms of the second law.

Now consider an inertial frame in which an isolated particle of constant *mass m* is subject to a *force* **F**. If one chooses one's units appropriately,[5] the

---

[4]Indeed, in his treatise *De Motu Corporum Gyrum* written in 1684, Newton postulated five laws of motion, one of which incorporated the Galilean principle of relativity. He later confined the dynamical laws to three, realising that these embodied all the salient features required for his theory.

[5]The actual equation is $\mathbf{F} = km\mathbf{a}$, where $k > 0$ is a universal constant. In SI units, force is measured in *newtons*. Defining $a = 1$ ms^{-2} when $F = 1$ N and $m = 1$ kg allows us to make $k = 1$.

second law can be defined as

$$\mathbf{F} = m\mathbf{a}, \tag{1.8}$$

where $\mathbf{a}$ is the acceleration of the particle due to $\mathbf{F}$. Alternatively, recalling that $\mathbf{a} = d\mathbf{v}/dt$,

$$\mathbf{F} = m\frac{d}{dt}\mathbf{v} \equiv m\dot{\mathbf{v}}$$
$$= \frac{d}{dt}(m\mathbf{v})$$
$$= \frac{d}{dt}\mathbf{p} \equiv \dot{\mathbf{p}},$$

where we have defined $m\mathbf{v} = \mathbf{p}$ as the *linear momentum* of the particle. The mass under discussion is actually the *inertial mass* of the particle. We will delay a formal account of the inertial mass until Section 1.3.

Newton's first two laws of motion explain the dynamical behaviour of isolated particles under the influence of external forces. One requires Newton's third law to understand how individual particles behave within a multi-particle system. We will resume discussion of N-particle systems in Chapter 6 and concentrate here on a two-particle system shown in Figure 1.7.

FIGURE 1.7
Two-particle system

Now, from Newton's second law, the force that acts on the ith particle $(i = 1, 2)$ is simply

$$\mathbf{F}_i = \dot{\mathbf{p}}_i. \tag{1.9}$$

This force is actually a combination of two forces: an *internal force* $\mathbf{F}_{ij}$, the effect that ith particle has on the jth particle, and an *external force*, $\mathbf{F}_i^{\text{ext}}$, that which may affect the system but in itself is not part of the system — for example, a gravitational force. Thence, the total force on each particle of our system is

$$\mathbf{F}_i = \mathbf{F}_i^{\text{ext}} + \sum_{j=1}^{2}\mathbf{F}_{ij}.$$

So, for particle P_1

$$\mathbf{F}_1 = \mathbf{F}_1^{\text{ext}} + \mathbf{F}_{11} + \mathbf{F}_{12}$$
$$= \mathbf{F}_1^{\text{ext}} + \mathbf{F}_{12} \tag{1.10}$$

and for particle P_2

$$\mathbf{F}_2 = \mathbf{F}_2^{\text{ext}} + \mathbf{F}_{21} + \mathbf{F}_{22}$$
$$= \mathbf{F}_2^{\text{ext}} + \mathbf{F}_{21}. \tag{1.11}$$

Notice that because no single particle can exert a force on itself, $\mathbf{F}_{ii} = \mathbf{0}$. Moreover, from Newton's third law, the force $\mathbf{F}_{12}$ that particle P_2 exerts on particle P_1 is equal and opposite to the force $\mathbf{F}_{21}$ that particle P_1 exerts on P_2:

$$\mathbf{F}_{12} = -\mathbf{F}_{21}. \tag{1.12}$$

On combining the definition of Newton's second law, (1.9), with (1.10) and (1.11) yields

$$\dot{\mathbf{p}}_1 = \mathbf{F}_1^{\text{ext}} + \mathbf{F}_{12}$$

and

$$\dot{\mathbf{p}}_2 = \mathbf{F}_2^{\text{ext}} - \mathbf{F}_{12},$$

where Newton's third law in the form of (1.12) has also been employed. Hence,

$$\dot{\mathbf{p}}_1 + \dot{\mathbf{p}}_2 = \mathbf{F}_1^{\text{ext}} + \mathbf{F}_2^{\text{ext}}.$$

If there are no net external forces then the total rate of change of momentum of the system, $\dot{\mathbf{p}} = \dot{\mathbf{p}}_1 + \dot{\mathbf{p}}_2$, is zero, and implies that the total momentum of the system is constant. This result is very important in its own right and is known as *the conservation of linear momentum*.

In particular, the forces $\mathbf{F}_{12}$ and $\mathbf{F}_{21}$ depicted in Figure 1.7 are called *central forces* — they act along the line of centre that joins P_1 and P_2. Although it is not a requisite property for the forces within a system to be central for Newton's third law to hold, invariably this will be the case for the majority of forces here encountered.

1.3 Gravitational and inertial mass

In our discussion so far, on the dynamical behaviour of *massive particles* (particles possessing mass), the 'mass' in question has been the *inertial mass* of the particle. So, for example, the mass appearing in Newton's second law, (1.8), is the inertial mass. In fact, one can define a further two distinct types of Newtonian mass that arise from his gravitational theory. They are defined as follows:

- *active gravitational mass*, m^{A}, is the measure of the matter producing the gravitational field;

- *passive gravitational mass*, m^{P}, is the measure of the matter subject to the gravitational field.

The equality of all three types of mass leads to the *weak equivalence principle*.

1.3.1 The weak equivalence principle

Following on from the meticulous astronomical observations of the Danish astronomer Tycho Brahe (1546-1601) and the subsequent analyses of these observations performed by Brahe's pupil Johannes Kepler (1571-1630), Newton formulated the first mathematical theory that not only was capable of explaining the process of falling apples, but could also yield a quantitative description of planetary motion in the solar system. The theory took the form of an (attractive) *inverse square law* and is known as *Newton's law of gravitation*.[6] The law asserts that *any two particles in the universe attract each other with a force that is proportional to the product of their masses and inversely proportional to the square of the distance between them*. The magnitude of this force is

$$\frac{Gm_1 m_2}{r^2},\tag{1.13}$$

where m_1 and m_2 are the masses of the particles, r is the distance between them and G is a constant known as the *gravitational constant*. The value of G accurate to three decimal places is 6.673×10^{-11} Nm^{2}kg^{-2}.

Now, the force acts along a line of centre that joins m_1 and m_2 and is also *conservative* (the energy of the system is conserved; this will be discussed in detail in Chapter 2). Hence, the *gravitational force* is a central, conservative force. Forces of this kind are dependent only upon the relative positions of the particles being acted upon.

With reference to Figure 1.8, the gravitational force $\mathbf{F}_{12}$ acting on P_1 (with mass m_1) due to P_2 (with mass m_2) is, by Newton's law of gravitation,

$$\mathbf{F}_{12} = \frac{Gm_1 m_2}{|\mathbf{r}_2 - \mathbf{r}_1|^3}(\mathbf{r}_2 - \mathbf{r}_1).\tag{1.14}$$

Conversely,

$$\mathbf{F}_{21} = \frac{Gm_2 m_1}{|\mathbf{r}_1 - \mathbf{r}_2|^3}(\mathbf{r}_1 - \mathbf{r}_2).\tag{1.15}$$

The acceleration of P_1 in (1.14) is $\ddot{\mathbf{r}}_1$. So, using Newton's second law,

$$\mathbf{F}_{12} = m_1^{\mathrm{I}}\ddot{\mathbf{r}}_1,$$

---

[6]There is some debate as to the true proponent of this law. It has been widely suggested that Robert Hooke (1635-1703) was responsible for postulating the idea in 1679, and that Newton suppressed Hooke's prior claim.

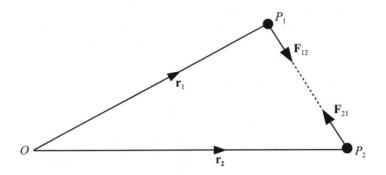

FIGURE 1.8
Newton's third law

where $m_1^{\rm I}$ is the inertial mass of P_1. Similarly,

$$\mathbf{F}_{21} = m_2^{\rm I}\ddot{\mathbf{r}}_2,$$

where $m_2^{\rm I}$ is the inertial mass of P_2. Appealing to the definition of active and passive gravitational mass above, equations (1.14) and (1.15) can now be written respectively as

$$m_1^{\rm I}\ddot{\mathbf{r}}_1 = \frac{Gm_1^{\rm P}m_2^{\rm A}}{|\mathbf{r}_2 - \mathbf{r}_1|^3}(\mathbf{r}_2 - \mathbf{r}_1) \tag{1.16}$$

and

$$m_2^{\rm I}\ddot{\mathbf{r}}_2 = \frac{Gm_2^{\rm P}m_1^{\rm A}}{|\mathbf{r}_1 - \mathbf{r}_2|^3}(\mathbf{r}_1 - \mathbf{r}_2).$$

However, applying Newton's third law (1.12) to these equations yields

$$m_1^{\rm P}m_2^{\rm A}(\mathbf{r}_2 - \mathbf{r}_1) = -m_2^{\rm P}m_1^{\rm A}(\mathbf{r}_1 - \mathbf{r}_2)$$
$$= m_2^{\rm P}m_1^{\rm A}(\mathbf{r}_2 - \mathbf{r}_1).$$

Hence,

$$\frac{m_1^{\rm P}}{m_1^{\rm A}} = \frac{m_2^{\rm P}}{m_2^{\rm A}}.$$

So, the passive/active gravitational mass ratios are identical for each particle in the system. On appropriately re-scaling the gravitational constant G, each ratio can be reduced to unity. One can therefore conclude that the active gravitational mass and passive gravitational mass are identical for each and every particle in a given system of particles.

There is no particular reason for the inertial mass and gravitational mass of a particle to be identical, other than experiment has established this to be true to a high degree of accuracy. Indeed, it is almost counterintuitive to assume that the inertial mass of a particle (a quantity defined in such a way as to have no direct connection with the force of gravity) should be responsible for the gravitational force between particles. Nevertheless, this is the case.

Newton was an ardent believer in the equality of gravitational and inertial mass. He conducted experiments with swinging pendulums composed of different materials. By observing that the periods of these pendulums remained the same regardless of the material composition, Newton concluded that within an experimental error 1 part in 10^3, the inertial/gravitational mass ratio was equal to one for a suitable choice of units. The equality of gravitational and inertial mass is known as the *weak equivalence principle*.

Consider (1.16) with a unique gravitational mass, $m^G = m^A = m^P$:

$$\ddot{\mathbf{r}}_1 = \frac{m_1^G}{m_1^I} \frac{Gm_2^G}{|\mathbf{r}_2 - \mathbf{r}_1|^3}(\mathbf{r}_2 - \mathbf{r}_1).$$

If $m_1^G/m_1^I = 1$, as experiment suggests, then the acceleration of P_1, due to P_2, is independent of the mass of P_1, and therefore constant for *any* particle located at the same point. Hence, letting the mass $m_2^G = M$ (the mass of the Earth) be positioned at the origin so that $\mathbf{r}_2 = \mathbf{0}$ and $\mathbf{r}_1 = \mathbf{r}$ the above equation reduces to

$$\mathbf{g} = -\frac{GM}{r^2}\hat{\mathbf{r}}, \qquad (1.17)$$

where $\hat{\mathbf{r}}$ is a unit vector in the direction of $\mathbf{r}$ (pointing radially out from the Earth's centre) and $\mathbf{g}$ is the *gravitational acceleration*. This result is a direct consequence of the weak equivalence principle.[7]

Following on from Newton, many other experimenters have made considerable strides in reducing the experimental error in establishing the equality between inertial and gravitational mass. Baron Loránd Eötvös de Vásárosnamény (aka Eötvös) (1848-1919) constructed a torsion balance by means of which he reduced the error to 5 parts in 10^9, and in 1972, Vladimir Braginsky and Vladimir Ponov refined the torsion balance experiment, yielding an error of 1 part in 10^{12}. At the time of writing, the most accurate results were those made by Stefan Baessler *et al.* in 1999, who found an error of 5 parts in 10^{14}. Other non-torsion balance experiments are currently being devised that if successful, could reduce the error to as little as 1 part in 10^{17} or even much less.

---

[7]Einstein assumed without reservation that inertial mass and gravitational mass were equivalent without error. Indeed, if this were not true the consequences would be catastrophic for general relativity.

1.4 Exercises

1. Consider a point P travelling along the x-axis. Its rectilinear motion is governed by
$$\ddot{x}(t) - C\dot{x}(t)^2 = 0,$$
where C is a non-zero constant with dimensions $(\text{Length})^{-1}$. Given initial conditions $\dot{x}(0) = v_0 > 0$ and $x(0) = x_0$, find the velocity and position of P at later t.

2. Consider a point P travelling along the x-axis. Its rectilinear motion is governed by
$$\ddot{x}(t) - 3\dot{x}(t) + 2x(t) = 0.$$
Given the initial conditions $\dot{x}(0) = v_0$ and $x(0) = x_0$, find the velocity and position of P at later t.

3. A particle is projected vertically in the positive $y(t)$ direction at ground level. Given that it is projected at time $t = T_0$ and attains a height $H = y(T)$, find the initial speed of projection and the greatest height attained by the particle.

4. A car travelling at 20 ms^{-1} has to apply its brakes at some instant in time. This produces a constant deceleration of 2 ms^{-2}. Find the distance that the car has covered before coming to rest.

5. A car travels a total distance of 126 m. Initially the car's speed is 6 ms^{-1}. During the first part of the journey, which covers 96 m, the car undergoes a uniform acceleration. During the second and final part of the journey the car undergoes a uniform deceleration before coming to rest. If the speed of the car is $v \text{ ms}^{-1}$ as it begins to decelerate, find the time taken for the car to cover each part of its journey. Your answer should be in terms of v alone.

   Find numerical values for v, the car's acceleration and the deceleration if the time taken to cover the total distance is 18 seconds.

6. An aeroplane begins to travel due north at 100 kmh^{-1}. A wind then blows in a northwesterly direction. What is the new velocity of the aeroplane?

7. John is riding a bicycle due east at 6 ms^{-1} while Mary is riding a bicycle S30°W at 10 ms^{-1}. What is the magnitude and direction of John's speed relative to Mary's?

8. A cat runs due east at 10 kmh^{-1} and is 50 m due north of a lazy dog. The dog attempts to catch the cat, but is not prepared to run any faster than 6 kmh^{-1}. Show that the lazy dog can come within no less than 40 m of the cat.

9. Prove that Newton's dynamical laws are invariant under Galilean transformations.

10. The wave equation for an electric field E is given by

$$\frac{\partial^2 E}{\partial x^2} = \mu\epsilon \frac{\partial^2 E}{\partial t^2},$$

where μ and ϵ are constants called the permeability and permittivity that characterise the medium through which the field propagates. Show that this equation is *not* invariant under Galilean transformations.

2

Particle Dynamics in One Dimension

CONTENTS

Newton's second law of motion will be the primary motivating factor in this chapter. As we are dealing with single-particle motion, Newton's third law becomes redundant, and his first law reminds us that all motion takes place in an inertial frame.

2.1 Motion of a particle under a force

The *equation of motion* that governs a particle's path through space is Newton's second law:

$$m\ddot{x} = F \tag{2.1}$$

or, equivalently,

$$\dot{p} = F,$$

where $p = m\dot{x}$ is the linear momentum of a particle. In general the force acting on a particle could be a function of more than one variable; that is, there could be a dependency on velocity and time as well as position:

$$F = F(x, \dot{x}, t).$$

The corresponding equation of motion is then the most general kind of second-order differential equation that can describe all conceivable particle motion under the action of F. So, if the position $x(t_0)$ and the velocity $\dot{x}(t_0)$ of a particle are known at a certain *initial time* $t = t_0$, the dynamical behaviour of the particle is then completely determined for all time $t > t_0$ and, in principle, for $t < t_0$. In essence, solving (2.1) will yield a unique solution provided that two initial conditions are given.

Example 2.1 A particle of mass m is constrained to move along the x-axis and is subject to a constant force k. Given initial conditions $\dot{x}(0) = v_0$ and $x(0) = x_0$, find the position of the particle at any subsequent time t.

Solution The equation of motion (2.1) is

$$\ddot{x} = \frac{k}{m}.$$

Integrating once yields the velocity of the particle

$$\dot{x}(t) = \int \ddot{x}(t)dt = \frac{k}{m}t + C_1,$$

where C_1 is an arbitrary constant of integration. The first initial condition fixes $C_1 = v_0$. Integrating a second time yields the position of the particle

$$x(t) = \int \dot{x}(t)dt = \frac{k}{2m}t^2 + v_0 t + C_2.$$

The second initial condition fixes $C_2 = x_0$. So, the position of the particle at any time t is

$$x(t) = \frac{k}{2m}t^2 + v_0 t + x_0. \qquad \square$$

Example 2.2 A particle of mass m is constrained to move along the x-axis and is subject to a force kx for $k > 0$. Given the boundary condition $\dot{x}(0) = v_0$, find the velocity of the particle at $x = x_0$.

Solution The equation of motion (2.1) is

$$\ddot{x} = \frac{k}{m}x$$

or, equivalently, with $\dot{x} = v$

$$\frac{dv}{dt} = \frac{k}{m}x.$$

It will be useful to consider the velocity of the particle as a function of displacement rather than time as we have our boundary conditions in terms of position. Using the chain rule, we have

$$\frac{dv}{dt} = \frac{dx}{dt}\frac{dv}{dx} = v\frac{dv}{dx}.$$

So, the equation of motion becomes

$$v\frac{dv}{dx} = \frac{k}{m}x.$$

This is a first-order separable equation with solution

$$\frac{v^2}{2} = \frac{k}{m}\frac{x^2}{2} + C.$$

On employing the boundary condition, the constant of integration is $C = v_0^2/2$. Hence, at $x = x_0$

$$v^2 = \frac{k}{m}x_0^2 + v_0^2. \qquad \square$$

2.1.1   Potential energy

Consider the *rectilinear* (straight-line) motion of a particle acted upon by a force $F(x)$ that depends on the position x of a particle. In this case, the equation of motion (2.1) is

$$m\ddot{x} = F(x)$$

or

$$m\frac{d\dot{x}}{dt} = F(x).$$

Now, from the chain rule

$$\frac{d\dot{x}}{dt} = \frac{dx}{dt}\frac{d\dot{x}}{dx} = \dot{x}\frac{d\dot{x}}{dx}.$$

Hence,

$$m\dot{x}\frac{d\dot{x}}{dx} = F(x).$$

Integrating this equation yields

$$\int m\dot{x}d\dot{x} = \int F(x)dx + \text{constant}.$$

Let us now introduce a function $V(x)$ such that

$$F(x) = -\frac{dV}{dx} \qquad (2.2)$$

by definition. Thence,

$$\int m\dot{x}d\dot{x} = -\int dV + \text{constant}$$

or

$$\frac{1}{2}m\dot{x}^2 + V(x) = \text{constant}. \tag{2.3}$$

The quantity $V(x)$ is called the *potential energy*, whereas the quantity $\frac{1}{2}m\dot{x}^2$ is the *kinetic energy* and is conventionally denoted by T. The constant on the right is the *total energy* E. Thus, for a particle moving with rectilinear motion and acted upon by a force that depends only on position, the sum of the particle's kinetic energy and potential energy yields a *conserved* quantity called the total energy of the particle:

$$T + V(x) = E. \tag{2.4}$$

This equation is called the *conservation of energy law.*

One can see from (2.2) that the potential energy must always involve an additive constant. Thus

$$V(x) = -\int_{x_0}^{x} F(x')dx'$$
$$= -\int F(x)dx + C, \tag{2.5}$$

where C is the result of evaluating the integral at suitable x_0 and the integration variable x has been replaced by the *dummy* variable x' to avoid possible confusion with the limits of integration. On letting $x = x_0$

$$V(x_0) = 0.$$

Thus x_0 is the zero level of potential energy. The choice of the position of zero potential energy is purely arbitrary and will not affect the physics. It is the difference in potential energy that is important,

$$V(x) - V(x_0) = -\int_{x_0}^{x} F(x')dx',$$

and so without loss of generality, one would be at liberty to set $V(x_0)$ to zero.

Example 2.3 A particle of mass m is projected vertically upwards (positive y-axis) from an arbitrary point $y = y_0$. If the particle is subject to a gravitational force $F(y) = -mg$, and air resistance is ignored, find the potential energy of the particle if it reaches a height y.

Solution Using equation (2.5)

$$V(y) = \int_{y_0}^{y} mg\,dy' = mg(y - y_0).$$

But $y - y_0$ is merely the height, h say, of the particle above y_0. So the potential energy can be written as

$$V = mgh.$$

Notice that if the particle were allowed to drop from an arbitrary point $y = y_0$ then the potential energy would be negative:

$$V = -mgh. \qquad \square$$

2.1.2 Work

The *work done* δw by the *force field* $F(x)$ when a particle moves from position x to $x + \delta x$ is defined as

$$\delta w = F(x)\delta x.$$

So, if a particle P moves from a point x_1 to a point x_2 under the influence of a force $F(x)$, the work done can be calculated using the line integral

$$W = \int_{x_1}^{x_2} F(x)dx.$$

If the path of the particle P is given parametrically by $x = x(t)$, then the work done by $F(x)$ in moving P between times t_1 and t_2 is

$$W = \int_{t_1}^{t_2} F(x)\dot{x}dt. \qquad (2.6)$$

Notice that from our discussion, we can relate the work done by a force in moving a particle from a point $x(t_1)$ to a point $x(t_2)$ to the potential energy of that particle at $x(t_2)$ relative to $x(t_1)$:

$$W = -\int_{x(t_2)}^{x(t_1)} F(x)dx = -[V(x(t_1)) - V(x(t_2))].$$

This is true provided that the force $F(x)$ is conservative; that is, the integral is independent of the path from $x(t_1)$ to $x(t_2)$.

The quantity represented by the integrand of (2.6) is the *power* and is defined as the *rate of work done*:

$$\text{Power} \equiv \frac{dW}{dt} = F\dot{x}.$$

But, $F\dot{x} = m\ddot{x}\dot{x} = d(m\dot{x}^2/2)/dt = \dot{T}$. Hence, the power is also the rate of kinetic energy:

$$\dot{W} = \dot{T}.$$

The SI unit of work is the joule (J) and the SI unit of power is the watt (W); therefore 1 watt = 1 joule per second.

2.2   Potential energy diagrams

The conservation of energy equation (2.3) can be rearranged to yield the velocity of a particle as a function of position:

$$\dot{x}(x) = \pm \left(\frac{2(E - V(x))}{m} \right)^{\frac{1}{2}}. \tag{2.7}$$

The ambiguity in sign (yielding two equal and opposite roots) arises because the particle's velocity could be in the positive or negative sense, since considering energy alone will not determine the outcome. In simple one-dimensional cases, the sign can usually be determined unambiguously.

Integrating (2.7) yields

$$t - t_0 = \pm \left(\frac{m}{2} \right)^{\frac{1}{2}} \int_{x_0}^{x} \frac{dx'}{(E - V(x'))^{\frac{1}{2}}}, \tag{2.8}$$

where t_0 is a constant of integration and the integration variable x has been replaced by the *dummy* variable x' to avoid possible confusion with the limits of integration. Once the potential energy $V(x)$ is given, one can evaluate this integral, in principle, to give t as a function of x. On inverting, x can be given as a function of t. Of course, it is assumed that the initial conditions t_0 and x_0 have already been given.

Notice from (2.3) that $\frac{1}{2}m\dot{x}^2 \geq 0$ ($\frac{1}{2}m\dot{x}^2 = 0$ if and only if $\dot{x} = 0$). So, any particle motion is necessarily confined to those regions where $V(x) \leq E$. It should be clear that in a region where the kinetic energy is zero, we have $V(x) = E$.

> **Example 2.4** A stone of mass m is released from rest and allowed to fall down a deep well. If the stone begins to fall at time $t = 0$, assuming it is released at $y = 0$, find the depth the stone has fallen at any later time. (Ignore air resistance.)
>
> **Solution** The potential energy in this situation has effectively been determined in Example 2.1; that is
>
> $$V(y) = -mgy.$$
>
> Initially the stone is released at $y = 0$ from rest ($\dot{y} = 0$) implying that the total energy E is zero. Employing (2.8) then gives
>
> $$t = \sqrt{\frac{m}{2}} \int_0^y \frac{dy'}{(mgy')^{\frac{1}{2}}} = \sqrt{\frac{2y}{g}}.$$

On rearranging, we have

$$y = \frac{1}{2}gt^2$$

as required.   □

It will not always be possible to solve (2.8) in a straightforward way as detailed in Example 2.4. Difficulties can arise if one is faced with a complicated expression for the potential energy. Then the integral might not readily be evaluated without resorting to numerical or graphical means. Indeed, a graphical analysis of a particle's potential energy, without recourse to solving (2.8), can yield a satisfactory qualitative account of the kind of particle motion that is physically feasible.

Consider the highly idealised plot of potential energy, $V(x)$, against position x in Figure 2.1. What does this potential energy diagram tell us about a particle's motion within the various regions of the diagram?

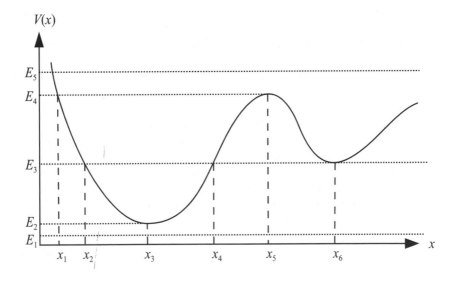

FIGURE 2.1
Potential energy diagram

To answer this question let us consider each of the energy levels in turn: E_1, E_2, etc.

For all values of x, $V(x) > E_1$. This implies that the velocity of a 'particle' travelling through any region along the x-axis will be imaginary. Such particles do not exist in classical dynamics; consequently, particles with energy E_1 are forbidden in our scenario.

A particle with total energy E_2 must reside indefinitely at x_3. At this point $V(x_3) = E_2$, which implies that the corresponding potential energy of the particle must be zero and motion cannot take place.

A particle with total energy E_3 may move within the region $x_2 \leq x \leq x_4$ or, for reasons stipulated in the previous case, reside indefinitely at x_6. Exactly what kind of motion occurs in the region $x_2 \leq x \leq x_4$? Consider, for a moment, our particle located at x_4. Here, the potential energy and total energy are equal: $V(x_4) = E_3$. Thus there is no kinetic energy and the particle is instantaneously at rest. However, the gradient of the potential curve, $dV(x)/dx$, is positive at x_4 and therefore non-zero. This means that, on referring to (2.2), there exists a force

$$F(x_4) = -\left(\frac{dV(x)}{dx}\right)_{x=x_4}$$

that tends to accelerate the particle along the negative x-direction. As the particle approaches x_3 it decelerates, kinetic energy increases and potential energy decreases. At x_3 the particle has zero acceleration, kinetic energy is at a maximum while potential energy is at a minimum — the particle continues along its path and approaches x_2. As the particle passes x_3 the gradient of the potential curve begins to increase negatively. On reaching x_2 the particle is again instantaneously at rest and a force $F(x_2)$ now acts to accelerate the particle in the positive x-direction. The entire process above is repeated provided there is no loss of total energy. The points x_2 and x_4 are called *turning points* and the repeated oscillation performed by the particle in this region is known as *periodic motion*.

A particle with total energy E_4 may move within the region $x_1 \leq x \leq x_5$. If the particle is initially located at x_1 a force $F(x_1)$ will act on the particle and accelerate it along the positive x-direction. Again the particle will pass the point at x_3 and continue on towards x_5. However, at the point x_5 the tangent to the potential energy curve is parallel to the x-axis. Thus $(dV/dx)_{x=x_5} = 0$, and there will be no force accelerating the particle back down the potential energy slope such that periodic motion can take place. Moreover, the particle located at x_5 has zero kinetic energy and therefore no velocity, because $V(x_5) = E_4$. So, in principle, the particle will reside at x_5 indefinitely. In practice, however, the particle will never reach x_5 in a finite time. Both the velocity and acceleration of the particle diminishes as it gets closer to x_5, but will be zero only at x_5 and this will take an infinite time to achieve.

A particle with a total energy $E_5 > E_4$ will have *unbounded* motion in the positive x-direction. This particle can come from a large distance x and negotiate the peak at x_5 and troughs at x_6 and x_3. As it reaches the point corresponding to total energy E_5 it will be instantaneously at rest before moving off in the positive x-direction, again without hindrance.

The points x_3, x_5 and x_6 have special names associated with them; they are called *equilibrium points*. These are points at which a particle can reside indefinitely. A further distinction can be made if one considers points where the potential energy curve has a local minimum or a local maximum. These particular points are categorized as *stable equilibrium points* and *unstable equilibrium points*, respectively.

A particle positioned at x_3 will undergo periodic motion in the *neighbourhood* of x_3 given a sufficiently small *impulse*.[1] This impulse[2] gives the particle a kinetic energy such that the combined total energy is $E_2 + \Delta E_2$, where ΔE_2 is very small. As this tiny additional energy will allow the particle only to oscillate in the neighbourhood of x_3, the point x_3 is an example of a stable equilibrium point. A similar argument can be made for x_6.

A particle positioned at x_5 will *not* undergo periodic motion in the neighbourhood of x_5 given an impulse of *any* magnitude. The point x_5 is an example of an unstable equilibrium point. Note that in the exceptional case of the potential energy curve having a horizontal point of inflexion, a particle displaced from this point would not undergo periodic motion in the neighbourhood of this point. Consequently horizontal points of inflexion are also unstable equilibrium points.

Acknowledging that equilibrium points can be identified by

$$\left(\frac{dV(x)}{dx} \right)_{x=x_0} = 0,$$

where the derivative is again evaluated at x_0 (the point of equilibrium), sufficient conditions for x_0 to be a stable or unstable equilibrium point are

$$\left(\frac{d^2 V(x)}{dx^2} \right)_{x=x_0} > 0$$

or

$$\left(\frac{d^2 V(x)}{dx^2} \right)_{x=x_0} < 0,$$

respectively. Provided that $(d^2 V(x)/dx^2)_{x=x_0} \neq 0$ these conditions are also necessary.

Example 2.5 A unit mass particle constrained to motion along the x-axis is subject to a force

$$F(x) = \frac{10(1-x)}{x^3}, \ x > 0.$$

Find the potential energy of the particle and draw the corresponding potential energy diagram. Indicate on your diagram the region where periodic motion may take place and the region in which the motion is

---

[1] Consider two particles P_1 and P_2 travelling with momentum $\mathbf{p}_1$ and $\mathbf{p}_2$ at times t_1 and t_2, respectively. The impulse of the force $\mathbf{F}$ (not necessarily conservative) is given by the integral $\int_{t_1}^{t_2} \mathbf{F} dt = \mathbf{p}_2 - \mathbf{p}_1$.

[2] If the impulse is too large the particle could gain enough potential energy and escape along the positive x-direction.

unbounded. Show, also, that there exists only one equilibrium point and that it is stable.

Solution The potential energy $V(x)$ is given by

$$V(x) = -\int F(x)dx$$

$$= -10 \int \frac{(1-x)}{x^3}dx, \quad x > 0$$

$$= \frac{5}{x^2} - \frac{10}{x},$$

where the constant of integration has been put to zero without loss of generality. The diagram representing this function is depicted in Figure 2.2.

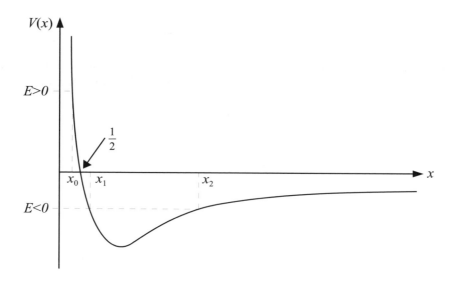

FIGURE 2.2
$V(x) = \frac{5}{x^2} - \frac{10}{x}$

For $0 < x_0 < 1/2$, $E > 0$ the particle's motion is unbounded and escapes to infinity. For $1/2 < x_0 < \infty$ the particle will oscillate between the turning points x_1 and x_2 if the total energy is negative $(E < 0)$. Only if $E > 0$ is unbounded motion possible. The particle can travel from $+\infty$ along the x-axis and come instantaneously to rest at x_0 where $0 < x_0 < 1/2$. The force acting on the particle at this point propels it again to $+\infty$ along the x-axis.

The equilibrium points are found using

$$\frac{dV(x)}{dx} = -F(x) = 0.$$

Thus,

$$-\frac{10(1-x)}{x^3} = 0 \implies x = 1.$$

So, there exists only one equilibrium point, which can also be seen from Figure 2.2.

$$\frac{d^2V(x)}{dx^2} = -\frac{20}{x^3} + \frac{30}{x^4} \quad \therefore \quad \frac{d^2V(1)}{dx^2} = 10 > 0.$$

Hence, the equilibrium point is indeed stable. ◻

Example 2.6 A particle of mass m is acted on by a force with potential energy

$$V(x) = 3x^2 - x^3.$$

Draw the potential energy diagram. Given the condition $\dot{x}(x) = v_0$ at $x = 0$, show that the particle is unable to escape to positive infinity if the magnitude of its velocity is

$$|v_0| \leq 2\sqrt{\frac{2}{m}}.$$

Solution The potential energy diagram for $V(x) = 3x^2 - x^3$ is given in Figure 2.3.

Now, the motion of the particle will be confined to a region $V(x) \leq E$. For the particle to escape to infinity it must possess sufficient kinetic energy $(T > 0)$ so that it can surmount the peak at $x = 2$. To prevent this occurring, T must be zero at $x = 2$. Hence,

$$E \leq V(2) = 4.$$

At $x = 0$, the potential energy $V(0) = 0$. Therefore,

$$E = T = \frac{1}{2}m\dot{x}(0)^2 = \frac{1}{2}mv_0^2.$$

Hence, for the particle *not* to escape to positive infinity

$$\frac{1}{2}mv_0^2 \leq 4$$

or

$$|v_0| \leq 2\sqrt{\frac{2}{m}}. \quad ◻$$

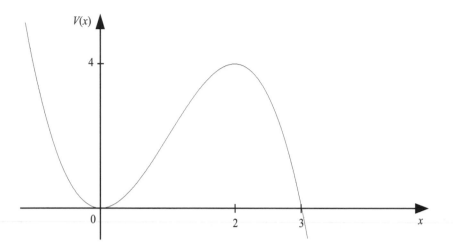

FIGURE 2.3
$V(x) = 3x^2 - x^3$

2.3   Tension

Consider the motion of a simple pulley system shown in Figure 2.4. The system consists of two particles connected by a string that passes over a fixed pulley. We will assume that the pulley is *smooth* so that the *tension* in the string remains constant along its length, otherwise the tensions in the string either side of the pulley would be imbalanced. The string itself will be considered as *light* and *inextensible*. Without these highly idealised properties, practical calculations would be extremely awkward.

By *light*, we mean that the string has no discernible mass and is infinitely thin. Thus the only force acting in the string will be the tension, which will be uniform throughout its length. The length of a taut inextensible string will remain unaffected under tension. This means that the magnitude of the acceleration of the attached particles will be equal under a gravitational force. Furthermore, each particle will be displaced by the same amount and their instantaneous velocities will be equal, at any given time.

To illustrate the dynamical behaviour of such a system, consider two masses m_1 and m_2, connected by a light inextensible string that is allowed to hang over the smooth fixed pulley in Figure 2.5.

Assume that $m_2 > m_1$ and the distance between m_2 and the point where the string just leaves the pulley is x. As this distance now determines the

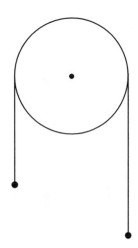

FIGURE 2.4
Simple pulley system

positions of m_1 and m_2 the velocity of each mass will be

$$v = \dot{x},$$

where v is chosen to be the upward positive velocity of m_1 and the downward positive velocity of m_2. It is necessary for a proper analysis to determine the forces acting separately on each mass. Thus, we need to establish the equation of motion for each mass in turn. Hence, if air resistance is ignored, the forces acting on m_1 and m_2 are

$$m_1 \ddot{x} = f_T - m_1 g$$

and

$$m_2 \ddot{x} = -f_T + m_2 g,$$

respectively, where f_T is the tension in the string. Notice that the acceleration $\ddot{x}$ is the same in both equations. It is conventional for the forces on the left of each equation to be taken as positive if the velocity associated with each particle is positive.

We can now solve these equations simultaneously to find the acceleration of the masses:

$$a = \ddot{x} = \left(\frac{m_2 - m_1}{m_2 + m_1}\right) g$$

and the tension in the string:

$$f_T = \left(\frac{2 m_1 m_2}{m_1 + m_2}\right) g.$$

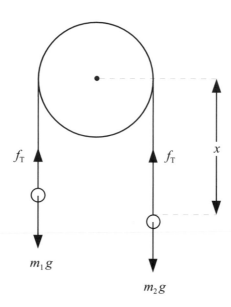

FIGURE 2.5
Simple pulley system with $m_2 > m_1$

Note that if m_2 were much larger than m_1, $a \approx g$ and $f_T \approx 2m_1g$.

2.4 Friction

Consider two bodies in static contact with each other; for example, a small block of wood of mass m sitting on a horizontal flat table (Figure 2.6).

The gravitational force acting on the block tends to push it into the table according to Newton's second law. If there were no other forces present, the block would indeed penetrate the surface of the table as easily as water through a sieve. However, classical physics forbids such behaviour, so there must exist some other force that prevents such an occurrence. Indeed, this force is called the *normal reaction* and acts perpendicularly to the plane of contact between the bodies.

Now, consider the table tilted by an angle θ to the horizontal (Figure 2.7). We can immediately see that this action induces additional forces, which makes modelling the dynamical behaviour of the block much more compli-cated. As well as the normal reaction **N** and the *weight* of the block $m\mathbf{g}$,

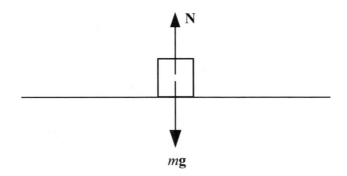

FIGURE 2.6
Two bodies in static contact

there are three other forces that must be taken into consideration in this situation.

The force represented by **f** is the *frictional force*. The laws governing frictional forces have been established empirically, so their occurrence is questionable. Nevertheless, we state them here for completeness.

Law of static friction: If two bodies are in contact without sliding taking place

$$\frac{f}{N} \leq \mu_{\mathrm{s}},$$

where μ_{s} is a number called the *coefficient of static friction*.

Law of kinetic friction: If two bodies are in contact and one slides over the other, which is at rest,

$$\frac{f}{N} = \mu_{\mathrm{k}},$$

where μ_{k} is a number called the *coefficient of kinetic friction*.

Both coefficients of friction depend only on the nature of the bodies. Their values are determined experimentally and fall within the following range:

$$0 \leq \mu_{\mathrm{k}} < \mu_{\mathrm{s}} < 1.$$

Going back to our block on an inclined table, if $\mu_{\mathrm{k}} = 0$ there is no frictional force and contact between the block and the table is defined as *smooth*. In this case, the slightest force acting on the block will cause it to slide down the table. For $\mu_{\mathrm{k}} \neq 0$ there is a frictional force and the contact between the block and the table is defined as *rough*.

The force **R** is the *resultant reaction* that the table exerts on the block

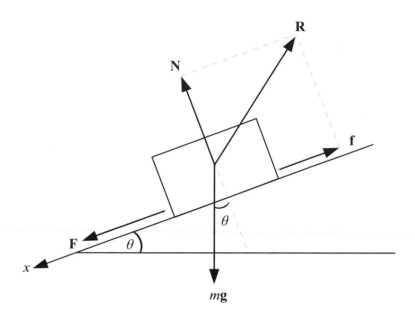

FIGURE 2.7
Block sliding down a table

and is written vectorially as
$$\mathbf{R} = \mathbf{N} + \mathbf{f}.$$
Provided that the block slides down the table in the positive x-direction without deviating in any other direction as indicated in Figure 2.7, the force $\mathbf{F}$ is by Newton's second law
$$\mathbf{F} = (m\ddot{x}, 0, 0).$$

Example 2.7 With reference to Figure 2.7, find the velocity of the block as it slides down the table at any time $t > 0$ given that it begins to accelerate from rest at $x = x_0$. Also, find the distance covered by the block in this time.

Solution Assuming that the block slides down the table in the positive x-direction, and does not deviate, we can resolve the weight of the block, $m\mathbf{g}$, in directions parallel and perpendicular to the x-axis. Thus, for the block to remain in equilibrium
$$m\ddot{x} = mg\sin\theta - f$$
and
$$N = mg\cos\theta.$$

But $f/N = \mu_k$; therefore,

$$m\ddot{x} = mg \sin\theta - \mu_k mg \cos\theta.$$

So the block slides down the table at a constant acceleration. We can now integrate this equation directly with respect to time and divide through by m, yielding

$$\dot{x}(t) = gt(\sin\theta - \mu_k \cos\theta) + C,$$

where C is a constant of integration. The given initial condition $\dot{x}(0) = 0$ implies that $C = 0$. Integrating again gives the distance covered by the block:

$$x(t) - x_0 = \frac{1}{2}gt^2(\sin\theta - \mu_k \cos\theta),$$

where the initial condition $x(0) = x_0$ has been incorporated directly.

Notice that if the surface were smooth $\mu_k = 0$ and the distance travelled by the block would be

$$x(t) - x_0 = \frac{1}{2}gt^2 \sin\theta. \qquad \square$$

2.5   Resistive motion

The upwards and downwards motion of a particle in the presence of a gravitational field was discussed in Examples 2.3 and 2.4. In both of these cases, we neglected to incorporate an extremely important facet of the particle's motion, that of air resistance. Indeed, all bodies that move through a *fluid* (either liquid or gas) will be subject to a resistive force referred to as *drag*. This resistive force can be most readily felt when swimming or riding a bicycle. Increasing one's speed through water or air will produce a larger drag effect, which inhibits the process of going faster. For some bodies, notably aeroplanes, there exists an additional force that acts in a direction perpendicular to their motion. This force is called *lift* and is the mechanism for flight. Our analysis, however, will be confined to bodies for which lift is negligible.[3]

Although drag is a function only of a body's velocity, the form of this resistive force will depend on such factors as the geometry of the body and the *density* and *viscosity* of the fluid. For a body travelling at low speed

---

[3] *Turbulence* can also play a crucial role, but for sufficiently low speeds, this too can be neglected.

through a fluid of high viscosity, the dynamical behaviour can be modelled
with the aid of the *linear drag law*:

$$f_l = k_l v,$$

where f_l is the linear drag force, v is the speed of the body and k_l is the *linear
drag coefficient*. The value of k_l will vary depending on the shape of the body.
For a sphere of radius a moving through a fluid of viscosity η the calculated
value of k_l is

$$k_l = 6\pi a \eta.$$

At higher, but subsonic, speeds and moving through a fluid of low viscosity,
the same body will have its dynamical behaviour modelled with the aid of the
quadratic drag law:

$$f_q = k_q v^2,$$

where f_q is the quadratic drag force, v is the speed of the body and k_q is
the *quadratic drag coefficient*. Again, the value of k_q will vary depending on
the shape of the body, but for the sphere above it has been experimentally
determined as

$$k_q = \frac{4}{5}\rho a^2,$$

where ρ is the density of the fluid.

The ratio of quadratic and linear drag forces incorporates a rather important dimensionless quantity called the *Reynolds number*: R_e. The Reynolds
number can quantitatively determine which of the drag laws is the most appropriate for a given physical situation. Thus,

$$\frac{f_q}{f_l} = \frac{k_q}{k_l}v = \frac{1}{15\pi}R_e,$$

where $R_e = 2\rho\eta^{-1}av$.

As an illustration, consider a baseball or cricket ball of radius 35 mm
travelling through the air at a speed of 10 ms^{-1}. If the density and viscosity
of air (at STP) are given as 1.2 kgm^{-3} and 1.7×10^{-5} m^{2}s^{-1}, respectively,
then

$$\frac{f_q}{f_l} \approx \left(\frac{2 \times 1.2 \times 0.035}{15\pi \times 1.7 \times 10^{-5}} \right) \times 10 \approx 1050.$$

The quadratic drag clearly dominates, and even if the speed of the ball were
reduced by 90 percent to 1 ms^{-1} this would still be the case. The corresponding
Reynolds number is

$$R_e = 15\pi \times 1050 = 49480,$$

which suggests that if the Reynolds number is large, the quadratic drag law
should be employed.

Suppose that now a small ball-bearing of radius 2 mm moves through castor oil at a speed of 0.1 ms^{-1}. Taking the density as 1000 kgm^{-3} and viscosity as 1.5 m^{2}s^{-1}, we have

$$\frac{f_q}{f_l} \approx \left(\frac{2 \times 1000 \times 0.002}{15\pi \times 1.5}\right) \times 0.1 = 0.002.$$

The quadratic drag is negligible compared to linear drag and the corresponding Reynolds number is

$$R_e = 15\pi \times 0.002 = 0.09.$$

This suggests that if the Reynolds number is low ($R_e \ll 1$) the linear drag law should be employed.

2.5.1   Vertical motion in a resistive medium

It has already been established above that one must take into consideration a number of different factors in determining which of the two drag laws will be most appropriate for a given physical situation. The value of the Reynolds number is a good qualitative indicator; however, for values approximately equal to unity, one must combine both linear and quadratic drag laws to obtain a more accurate model of the dynamical behaviour of the system. Unfortunately, the corresponding equation of motion becomes non-linear, so the solving process becomes much more difficult.

In the following analysis, only linear or quadratic resistance (never the combination) will be encountered. This makes the equations of motion readily soluble and the physics easier to understand.

Consider a projectile of mass m launched *vertically upwards* with speed u in a medium such that the drag force acting on the projectile has magnitude $k_1 v$ where $v = dy/dt$ is the speed at time t, and y is the vertical height above the ground ($y = 0$) (Figure 2.8).

When the projectile moves upwards the linear resistance acts downwards (always in the opposite direction to motion), and as gravity acts downwards by definition, the equation of motion, Newton's second law, is

$$m\frac{dv}{dt} = -mg - k_1 v$$

or

$$\frac{dv}{dt} + \frac{k_1 v}{m} = -g.$$

Our intent lies in establishing a form for the velocity and subsequently the maximum height attained by the projectile. Although the equation of motion is separable, it will be useful to use an integrating factor method to first obtain the velocity. Thus

$$\frac{d}{dt}(ve^{k_1 t/m}) = -ge^{k_1 t/m}$$

or, after integrating,

$$v(t) = -\frac{mg}{k_1} + C_1 e^{-k_1 t/m},$$

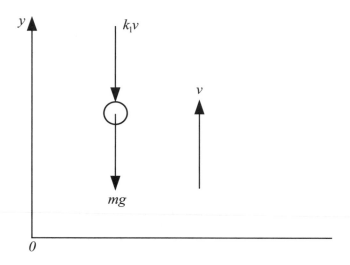

FIGURE 2.8
Upwards motion of a projectile with linear resistance

where C_1 is a constant of integration. Applying the initial condition $v(0) = u$ gives $C_1 = u + mg/k_1$. Therefore,

$$v(t) = \left(u + \frac{mg}{k_1}\right)e^{-k_1 t/m} - \frac{mg}{k_1}.$$

Notice that as $t \to \infty$, $v \to -mg/k_1$. This, as we will see for the case of a falling projectile, is the terminal velocity — the minus sign indicates downward motion. Also, as t increases finitely v decreases and when $v = 0$ the time taken for the projectile to reach its maximum height is

$$e^{k_1 t/m} = 1 + \frac{k_1 u}{mg}$$

or

$$t = \frac{m}{k_1} \ln\left(1 + \frac{k_1 u}{mg}\right).$$

To obtain the maximum height h of the projectile it will be convenient to express v as a function of y instead of t. This is because time will not feature in the final result. Thus

$$\frac{dv}{dt} = \frac{dv}{dy}\frac{dy}{dt} = v\frac{dv}{dy}.$$

So the equation of motion becomes

$$v\frac{dv}{dy} = -\left(g + \frac{k_1 v}{m}\right).$$

This equation is separable and so can be written as

$$\int dy = -\int \frac{v\,dv}{g + k_1 v/m}$$

$$= -\frac{m}{k_1} \int \left(1 - \frac{g}{g + k_1 v/m}\right) dv,$$

which implies

$$y = -\frac{m}{k_1}\left(v - \frac{mg}{k_1}\ln(g + k_1 v/m)\right) + C_2,$$

where C_2 is a constant of integration. Incorporating the conditions $v = u$ when $y = 0$ gives $C_2 = \frac{m}{k_1}\left(u - \frac{mg}{k_1}\ln(g + k_1 u/m)\right)$. Thus,

$$y = \frac{m}{k_1}(u - v) - \frac{m^2 g}{k_1^2}\ln\left(\frac{g + k_1 u/m}{g + k_1 v/m}\right).$$

The maximum height is attained when $v = 0$. Hence,

$$h = \frac{mu}{k_1} - \frac{m^2 g}{k_1^2}\ln\left(1 + \frac{k_1 u}{mg}\right).$$

The same projectile is now allowed to fall from rest vertically downwards through the same resisting medium (Figure 2.9). In this case the positive y-direction is taken downwards. In this direction the sign of mg is positive while that of $k_1 v$ is negative. Thus the equation of motion is written as

$$m\frac{dv}{dt} = mg - k_1 v \tag{2.9}$$

or

$$\frac{dv}{dt} + \frac{k_1 v}{m} = g.$$

The integrating factor method will again be used to solve this first-order equation. Multiplying through by the integrating factor $e^{k_1 t/m}$ gives

$$\frac{d}{dt}(v e^{k_1 t/m}) = g e^{k_1 t/m}.$$

On integrating,

$$v(t) = \frac{mg}{k_1} + C_3 e^{-k_1 t/m}.$$

Incorporating the initial conditions $v(0) = 0$ gives $C_3 = -mg/k_1$. Hence,

$$v(t) = \frac{mg}{k_1}(1 - e^{-k_1 t/m}). \tag{2.10}$$

Now as $t \to \infty$, $v(\infty) \equiv v_\infty \to mg/k_1$. This is called the *terminal speed* of the projectile and is achieved when the weight of the projectile and the

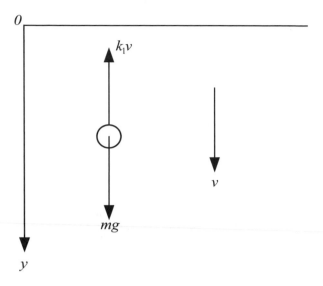

FIGURE 2.9
Downwards motion of a projectile with linear resistance

linear drag are in equilibrium: $mg = k_1 v$. Of course, when this occurs the acceleration of the projectile is zero, so the terminal speed could easily have been established by putting $\dot{v} = 0$ in (2.9).

In practice the projectile will never actually achieve its terminal speed, because it would take an infinite time to do so. However, we can show that a high percentage of a body's terminal speed will be reached in a short time. For example, let us calculate the time it would take for a projectile to reach 95 percent of its terminal speed. The terminal speed with linear drag is $v_\infty = mg/k_1$ and $v = 0.95 v_\infty$. Substituting these values into (2.10) yields

$$0.95 = 1 - e^{-gt/v_\infty}.$$

Rearranging gives

$$t = \frac{\ln 20}{g} v_\infty \approx 0.3 v_\infty.$$

For a projectile with a calculated v_∞ of, say, 3 ms^{-1}, it will take under a second for the projectile to reach 95 percent of its terminal speed.

For short durations of fall, the velocity of the projectile can be obtained by expanding the exponential term in (2.10):

$$v(t) = gt - \frac{1}{2}gt^2 \frac{k_1}{m} + \cdots.$$

So if $t \ll m/k_1$, $v \approx gt$ and the linear drag is negligible. Notice that because the terminal speed is proportional to m and $1/k_1$, projectiles of different sizes moving through a medium with linear drag k_1 will have different terminal speeds. Thus, the larger the mass, the higher the terminal speed. Also, larger linear drags correspond to lower terminal speeds. This is why differently sized bodies dropped from the Leaning Tower of Pisa, say, will *not* reach the ground simultaneously.

Bodies with large masses and terminal speeds that move through less viscous mediums, such as air, are better modelled by means of an equation of motion incorporating a quadratic drag force of magnitude $k_q v^2$, where again $v = dy/dt$. If the y-direction is taken vertically upwards (see Figure 2.10) then

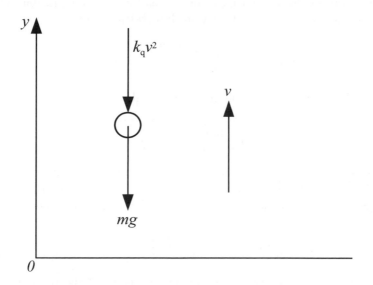

FIGURE 2.10
Upwards motion of a body with quadratic resistance

the equation of motion for a body projected vertically upwards is

$$m\frac{dv}{dt} = -mg - k_q v^2$$

or

$$\frac{dv}{dt} = -\left(g + \frac{k_q v^2}{m}\right).$$

If the body is projected with initial speed u, we can determine the maximum height h attained by the body in a similar way as that for linear drag. Thus, we must have v as a function of y:

$$\frac{dv}{dt} = \frac{dv}{dy}\frac{dy}{dt} = v\frac{dv}{dy} \equiv \frac{1}{2}\frac{d}{dy}(v^2).$$

The equation of motion can then be written as

$$\frac{d}{dy}(v^2) = -2\left(g + \frac{k_q v^2}{m}\right),$$

which gives

$$\int_{u^2}^{0} \frac{d(v^2)}{g + k_q v^2/m} = -\int_{0}^{h} 2dy,$$

where the conditions have been incorporated as limits of integration, which avoids the use of a constant of integration. Integrating yields

$$\frac{m}{k_q}\left[\ln\left(g + \frac{k_q v^2}{m}\right)\right]_{v^2=u^2}^{v^2=0} = -2h.$$

Substituting in the limits gives the maximum height as

$$h = \frac{m}{2k_q}\ln\left(1 + \frac{k_q u^2}{mg}\right).$$

Notice that as $k_q \to 0$, $\frac{m}{2k_q} \to \infty$, which seems to imply that in the absence of a frictional force the body will attain an infinite height! Clearly, this cannot be the case. This problem can be resolved if the log term is expanded:

$$h = \frac{m}{2k_q}\left(\frac{k_q u^2}{mg} - \frac{1}{2}\left(\frac{k_q u^2}{mg}\right)^2 + \cdots\right).$$

On considering the leading term, the maximum height of the body without quadratic drag is

$$h = \frac{u^2}{2g}.$$

Let us now consider the same body falling through the same resistive medium, and again let the upwards y-direction be positive (Figure 2.11). The sign of mg is again negative but the quadratic drag is now positive — opposite to the direction of motion.

The equation of motion is

$$m\frac{dv}{dt} = -mg + k_q v^2$$

or

$$\frac{dv}{dt} = -\left(g - \frac{k_q v^2}{m}\right).$$

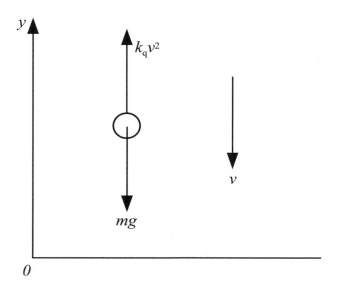

FIGURE 2.11
Downwards motion of a body with quadratic resistance

This can be written as

$$-\int_0^v \frac{dv'}{g - k_q v'^2/m} = \int_0^t dt',$$

where the limits of integration have been directly incorporated. Note that the dummy integration variables have been changed from v and t to v' and t' so as not to cause confusion with the limits of integration. The above integral can be evaluated using a natural log or hyperbolic tangent substitution. We will use the latter, but the reader may wish to try using the former as an exercise. Hence,

$$t = -\sqrt{\frac{m}{k_q g}} \tanh^{-1}\left(\sqrt{\frac{k_q}{mg}}\, v\right).$$

This can be solved for v, giving

$$v(t) = -\sqrt{\frac{mg}{k_q}} \tanh\left(\sqrt{\frac{k_q g}{m}}\, t\right).$$

As in the case of linear drag, the terminal speed of the body is obtained by letting $t \to \infty$ or putting $dv/dt = 0$. In either case, the terminal speed v_∞ is given by

$$v_\infty = \sqrt{\frac{mg}{k_q}}.$$

Note that the minus sign has been omitted in front of the square root to denote terminal speed and not terminal velocity, which is defined for a given direction: up or down.

2.6 Escape velocity

It is not easy to leave the surface of a planet unless, of course, you happen to be a crew member of the Starship Enterprise — and are not wearing red! The obstacle is the gravitational field of the planet, which can be overcome only by attaining a sufficiently high velocity — the *escape velocity*.

Suppose we wish to launch a body vertically upwards so that it just manages to escape the gravitational field of the Earth. What magnitude of escape velocity will be sufficient for this to occur? To answer this question, we begin by writing down Newton's law of gravity for the two masses involved; that is, the mass of the Earth M and the mass of our body m. Thus, with the aid of (1.13)

$$F = -\frac{mMG}{(R+y)^2},$$

where R is the radius of the Earth and y is the height attained by the body above the Earth's surface after it is launched at $y = 0$. The gravitational potential energy can be determined using (2.5),

$$V(y) = \int_{\infty}^{y} \frac{mMG}{(R+y)^2}\, dy = -\frac{mMG}{R+y}, \tag{2.11}$$

where the lower limit of infinity has been used to avoid incorporating a constant of integration.

The dynamical behaviour of the body is more easily understood if a potential energy diagram is plotted (Figure 2.12) for the function (2.11).

Now, any possible motion of the body is confined to regions where $V(y) \leq E$. This occurs when $E \geq 0$. For $E < 0$ the body will attain a certain height before falling back to Earth. Using conservation of energy (2.3), both the escape velocity and maximum height can be determined. For $E < 0$:

$$-\frac{mMG}{R+y} = E,$$

implying that the maximum height attained by the body above the Earth's surface is

$$y_{\text{max}} = -\frac{mMG}{E} - R.$$

Notice that there is no kinetic energy term in the conservation equation because the velocity of the body at y_{max} is necessarily zero.

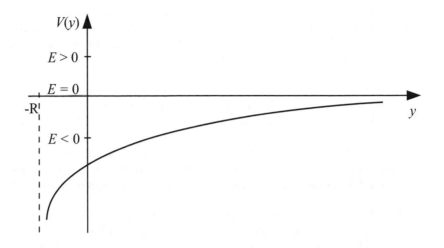

FIGURE 2.12
Potential energy diagram for $V(y) = -mMG/(R+y)$

For $E \geq 0$:

$$\frac{1}{2}m\dot{y}^2 - \frac{mMG}{R+y} \geq 0$$

implying that

$$|\dot{y}| \geq \sqrt{\frac{2MG}{R+y}}.$$

Thus, the minimum speed with which the body must be launched, at a height y *above* the Earth's surface, to escape the gravitational field is

$$\dot{y}_{\min} = \sqrt{\frac{2MG}{R+y}}.$$

If the body were launched *at* the Earth's surface $(y = 0)$ the speed v_{escape} necessary for escape is

$$v_{\text{escape}} = \sqrt{\frac{2MG}{R}}. \tag{2.12}$$

This result can be written in terms of g and R with the aid of (1.17). Thence

$$v_{\text{escape}} = \sqrt{2gR}.$$

This result is valid for any spherical body with known radius and gravitational acceleration. In particular, for the Earth, $g \approx 9.8 \text{ ms}^{-2}$ and $R \approx 6.37 \times 10^6$ m giving $v_{\text{escape}} = 11 \times 10^3 \text{ ms}^{-1}$.

As early as the 1770s, certain people were speculating about the possible existence of heavenly bodies whose gravitational fields were so strong that even light itself could not escape their surfaces. In particular, the Reverend John Michell (1724-1793) has been credited as giving the earliest analytical account of such phenomena. He writes:

> "If the semidiameter of a sphaere (sic) of the same density with the sun were to exceed that of the sun in the proportion of 500 to 1, a body falling from an infinite height towards it, would have acquired at its surface a greater velocity than that of light, and consequently supposing light to be attracted by the same force in proportion to its *vis inertiae* with other bodies, all light emitted from such a body would be made to return towards it, by its own proper gravity."[4]

Although a little tentative by today's standards, Michell reasoned that v_{escape} for the Sun was around $c/497$ (c being the velocity of light). He then argued that for a heavenly body with density equal to the Sun's and a radius approximately 497 times that of the Sun, v_{escape} would be greater than c. Therefore, light would be unable to escape its surface, rendering it invisible but not undetectable. In principle, one could detect its presence by the way in which other luminous bodies may orbit around such a *black hole*.

Example 2.8 Given that the mass of the Earth is 5.97×10^{24} kg and the gravitational constant $G = 6.67 \times 10^{-11}$ Nm^{2}kg^{-2}, determine the radius of the Earth if it were to transform into a black hole, assuming no change in mass takes place during the transformation.

Solution Substituting the given values into (2.12) and rearranging for R yields

$$R_{\text{black hole}} = \frac{2(5.97 \times 10^{24})(6.67 \times 10^{-11})}{(3 \times 10^8)^2} = 8.85 \times 10^{-3} \text{ m},$$

where $v_{escape} = c = 3 \times 10^8$ ms^{-1}. ☐

So unless there is some kind of planet-pounding giant lurking in the cosmos, Earth is safe from being squashed to a black hole the size of a blueberry!

2.7 Exercises

1. A force $F(t)$ acts on a particle of mass m such that its path is a straight line along the x-axis. Given that $F(t) = \sin \omega t$, where ω is

---

[4] "On the Means of Discovering the Distance..." Philosophical Transactions, lxxiv (1784), page 42

the constant angular frequency, find the position of the particle as a function of t.

2. A force $F(x) = -kx^2$ acts on a particle of mass m such that its path is a straight line along the x-axis and k is a positive constant. Obtain the total energy of the particle and draw the potential energy diagram. If the particle is moving with speed u in the positive x-direction at the point $x = 0$ show that there exists only one equilibrium point and find it.

3. A particle of mass m is acted on by a force with potential energy

$$V(x) = k\left(\frac{x}{x^2 + a^2}\right),$$

where k is a positive constant. Show that there exists only one equilibrium point and that it is stable. As the particle moves through the equilibrium point with speed u find the range of u for which the particle

(i) performs periodic motion

(ii) escapes to $+\infty$

(iii) escapes to $-\infty$.

4. For a certain diatomic molecule, the force between the nuclei of the two atoms has the following potential energy function:

$$V(x) = \frac{a}{x^{12}} - \frac{b}{x^6},$$

where x is the distance between the atoms and a and b are positive constants. Explain the significance of each term in this expression and draw a potential energy diagram that incorporates these terms individually and in combination. Assume that one of the atoms is at rest while the other is able to move with rectilinear motion. Show that there exists one equilibrium point and that it is stable.

5. A particle of mass 12 kg is connected to a light scalepan by a light inextensible string that passes over a smooth fixed pulley. Two cubes c_1 and c_2 are positioned on the scalepan such that c_1 rests directly on top of c_2. Given that the masses of c_1 and c_2 are 5 kg and 2 kg, respectively, find the acceleration of the system, the tension in the string and the normal reaction between the two cubes.

6. A particle A of mass 2 kg is connected to a particle B of mass 4 kg by a light inextensible string that passes over a smooth fixed pulley. Initially, the system is positioned such that both particles are at a height of 1 m above a table and then released from rest. Subsequently, particle B hits and coalesces with the table. Assuming that particle A never reaches the pulley during its upward motion, determine the maximum height attained by A.

7. A particle A of mass 4 kg is connected to a particle B of mass 6 kg by a light inextensible string that passes over a smooth fixed pulley. The pulley is attached to the top of a rough plane, which is inclined at an angle to the horizontal. Particle A is initially at rest on the plane while B is hanging freely. The coefficient of kinetic friction between A and the plane is $1/2$. Find:

    (i) the acceleration of the system after it is released from rest
    (ii) the tension in the string
    (iii) the force acting on the pulley.

8. A body is projected vertically downwards with speed u in a medium such that the drag force acting on the body has magnitude kv per unit mass. Show that as $t \to \infty$, $v \to g/k$. A similar body is now projected at speed u' in the same direction as the first body but t' seconds later. Show that

    $$\lim_{t \to \infty} \Delta y = (u - u' + gt')/k,$$

    where Δy is the distance between the two bodies.

9. Consider a body moving rectilinearly in a medium with a resistive force of magnitude mkv^{n+1}, where k is a positive constant and v is the speed of the body. If the speed is u at $t = 0$ show that at any later time

    $$\left(\frac{v}{u}\right)^n = \frac{1}{1 + nku^n t}, \quad n+1 > 0, \quad n \neq 0.$$

    If $-1 < n < 0$, show that the speed of the body becomes zero in a finite time. Show also that the body has travelled a distance

    $$\frac{(1 + nku^n t)^{(n-1)/n} - 1}{(n-1)ku^{n-1}}, \quad n \neq 1$$

    at a later time t.

10. A body of mass m moving rectilinearly is subject to a drag force

    $$F = -ke^{av},$$

    where k and a are positive constants and v is the speed. If the initial speed is u show that the body will come to rest in a time

    $$\frac{m}{ka}(1 - e^{-au})$$

    and at a distance

    $$\frac{m}{ka^2}(1 - (1 + au)e^{-au}).$$

11. A body falls vertically through a resistive medium with a resistive force of magnitude (i) kv, and (ii) kv^2. If the terminal speed of the body is v_T in both cases, show that the speed and distance covered by the body is, respectively,

(i) $v_T(1 - e^{-gt/v_T})$, $\quad v_T t - (v_T^2/g)(1 - e^{-gt/v_T})$

(ii) $v_T \tanh(gt/v_T)$, $\quad (v_T^2/g)\ln[\cosh(gt/v_T)]$.

3

Oscillations

CONTENTS

Oscillations, especially simple harmonic oscillations, are an extremely important part of dynamics and physics *per se*. They occur in everything from the balance of a wheel in a mechanical watch, to the molecules in a solid body. Moreover, there exist a large number of different dynamical problems that can be solved in a straightforward manner if the systems being investigated are linear.

3.1 Hooke's law

Consider a body attached to one end of a spring and at rest on a smooth horizontal table, as shown in Figure 3.1. As depicted, the spring is in a state of equilibrium; that is, the spring is neither being *compressed* nor *extended*. Assuming that motion can take place only in the x-direction, extending (or compressing) the spring by a small amount produces a *restoring force $F(x)$* that opposes the extension (or compression), as shown in Figure 3.2.

The restoring force is modelled by the equation

$$F(x) = -kx, \qquad (3.1)$$

where k is the positive *spring constant* and x is the displacement of the body from its equilibrium position. Notice that $F(x)$ is also conservative as the force depends only upon position. Equation (3.1) is known as *Hooke's Law* and was

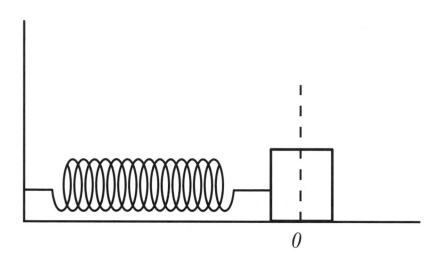

0

FIGURE 3.1
Body attached to a spring

discovered by Robert Hooke (1635-1703) in 1675 and published the following year.[1]

The dynamical behaviour of a body subject to Hooke's law can best be illustrated with the aid of a potential energy diagram (see Figure 3.3). The potential energy associated with Hooke's law is, by (2.5),

$$V(x) = \frac{1}{2}kx^2,$$

where the equilibrium position is at $x = 0$. We will shortly see that this must be a stable equilibrium as any *small* extension or compression of the spring will necessarily result in the body relocating back to the origin as in Figure 3.1.

Let us now consider the body in Figure 3.1 performing *small oscillations* about the equilibrium position $x = 0$. In such cases it is permissible to expand the potential energy using the Maclaurin series:

$$V(x) = V(0) + x\frac{dV(0)}{dx} + \frac{1}{2}x^2\frac{d^2V(0)}{dx^2} + \cdots.$$

The constant term $V(0)$ can be set equal to zero without affecting the physics

---

[1]So as to conceal salient features and thus keep his competitors in comparative darkness, Hooke published many of his scientific articles with anagrammatic titles; this also conjured an air of mystique, for which he had a penchant. The anagram used for the law that bears his name is "ceiiinosssttuv", which when rearranged, translates in Latin as "ut tensio sic vis" (the power of the spring is proportional to its tension).

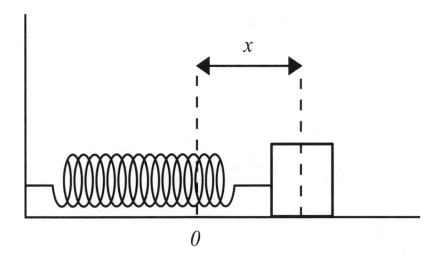

FIGURE 3.2
Body attached to a spring extended by a distance x

of the system. As $x = 0$ is the equilibrium position, the condition that satisfies this is $dV(0)/dx = 0$. Thus, for small oscillations about $x = 0$, we can neglect terms of order x^3 or higher, giving the potential energy as

$$V(x) = \frac{1}{2}kx^2, \tag{3.2}$$

where $k \equiv d^2V(0)/dx^2$. For $k > 0$, (3.2) is depicted by the parabola in Figure 3.3. Thus, the body's motion is confined to regions $V(x) \leq E$ with allowed total energies $E \geq 0$. If $E = 0$, there is no kinetic energy and the body will remain at rest in the equilibrium position $x = 0$. If $E > 0$ the body will perform small oscillations about $x = 0$ in the region $-x_0 \leq x \leq x_0$, coming to rest momentarily at $x = \pm x_0$ where x_0 is called the *amplitude*. In terms of total energy E and k, the maximum displacement of the body is

$$x = \pm\sqrt{\frac{2E}{k}}, \quad V(x) = E.$$

For $k < 0$, the equilibrium position at $x = 0$ is unstable and the force acting on the body will be repulsive rather than attractive. Such a case gives our current model an unphysical character and will not be considered here.

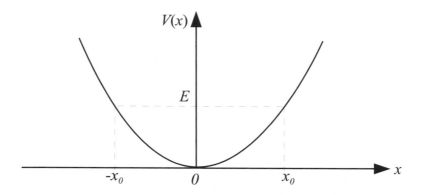

FIGURE 3.3
The potential energy for the case $k > 0$

3.2 Simple harmonic motion

The model given in the last section of a body attached to a spring performing small oscillations about an equilibrium point is an example of a *linear* or *harmonic oscillator*. This type of one-dimensional motion is called *simple harmonic motion*, and is the most fundamental of all the oscillations in physics — and fortunately, the easiest to solve. The equations of motion governing the various types of simple harmonic motion are ordinary linear differential equations with constant coefficients; and as we shall see, these are uncomplicated to solve. It should be mentioned that Hooke's law is a good model for springs over a limited range of extension and compression. If the body in Figure 3.1 were extended or compressed such that the ensuing oscillations were not considered "small", or equivalently k is a function of displacement, the motion would no longer be simple harmonic, but instead non-linear. Equations governing this type of dynamical behaviour are considerably harder to solve.

The equation of motion for the harmonic oscillator depicted in Figure 3.1 is, from Newton's second law,

$$m\ddot{x} = F(x) = -kx. \tag{3.3}$$

At this stage, we are assuming that no other forces are acting, such as damping, resistive and driving forces. It will be useful to rewrite this equation in the following form:

$$\ddot{x} + \omega^2 x = 0, \tag{3.4}$$

where $\omega = \sqrt{k/m}$ is called the *angular frequency* of the oscillations of the body about its equilibrium position.

This equation of simple harmonic motion is a linear second-order *homogeneous* differential equation with constant coefficients. Equations such as these that lack a function of the independent variable only, (in this case t) on the right-hand side, are said to be homogeneous.[2]

The function $x(t) = e^{\lambda t}$ is a solution of (3.4) provided that λ satisfies the *auxiliary equation*

$$\lambda^2 + \omega^2 = 0.$$

Thus $\lambda = \pm i\omega$, yielding two independent complex solutions

$$x_1(t) = e^{i\omega t}$$

and

$$x_2(t) = e^{-i\omega t}.$$

Since (3.4) is linear, it is permissible to construct a *linear combination* of the two independent solutions forming the *general solution*

$$x(t) = C_1 x_1(t) + C_2 x_2(t)$$
$$= C_1 e^{i\omega t} + C_2 e^{-i\omega t}, \tag{3.5}$$

where C_1 and C_2 are arbitrary *complex* constants. This form of solution is necessarily complex; however, to gain any physical meaning of the problem, the general solution must be real, which means that C_1 and C_2 must be chosen appropriately. To achieve real solutions, we must invoke *Euler's formula*:

$$e^{\pm i\omega t} = \cos \omega t \pm i \sin \omega t. \tag{3.6}$$

Substituting (3.6) into (3.5) yields

$$x(t) = (C_1 + C_2) \cos \omega t + i(C_1 - C_2) \sin \omega t.$$

So, x is real provided that $C_1 = C_2^*$ and $C_2 = C_1^*$, where (*) denotes complex conjugation, giving

$$x(t) = (C_1 + C_1^*) \cos \omega t + i(C_1 - C_1^*) \sin \omega t.$$

The coefficients of $\cos \omega t$ and $\sin \omega t$ are now necessarily real and can be replaced such that

$$x(t) = B_1 \cos \omega t + B_2 \sin \omega t, \tag{3.7}$$

where

$$B_1 = C_1 + C_1^* \quad \text{and} \quad B_2 = i(C_1 - C_1^*).$$

Thus, the *real* general solution of the simple harmonic motion equation is given by (3.7).

---

[2]Strictly speaking, a second-order linear equation of the form $a(t)\ddot{x}(t) + b(t)\dot{x}(t) + c(t)x(t) = f(t)$ (for functions $a(t)$, $b(t)$, $c(t)$ and $f(t)$) is called homogeneous if $f(t) = 0$ for all t.

The dynamical behaviour of the body in Figure 3.1 can most easily be analysed by rewriting (3.7) in a slightly different way:

$$x(t) = A\cos(\omega t - \phi), \qquad (3.8)$$

where we have employed the familiar trigonometric identity and set $B_1 = A\cos\phi$ and $B_2 = A\sin\phi$. The real constants A and ϕ will be defined presently.

The simple harmonic motion of the oscillating body in Figure 3.1 is depicted in Figure 3.4. If the body is extended (or compressed) by a small

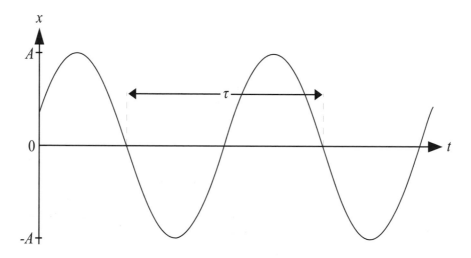

FIGURE 3.4
Simple harmonic motion $x(t) = A\cos(\omega t - \phi)$

amount, it will oscillate about its equilibrium position indefinitely — recall that the body sits upon a *smooth* table. The positive constant A is called the *amplitude* and is the maximum displacement of the body from its equilibrium position. The constant ϕ is the *phase shift* or *phase angle*; it ranges between 0 and 2π and defines the position of the body in oscillation at time $t = 0$. So the position from which the body commenced oscillating is

$$x = A\cos\phi.$$

If the body happened to commence oscillating from $x = 0$ at $t = 0$, then its displacement at a later time t is

$$x = A\cos\omega t.$$

In all other circumstances the motion is defined by (3.8). Notice that (3.8) is $(2\pi/\omega)$-*periodic* — the function $x(t)$ repeats itself after a time

$$\tau = \frac{2\pi}{\omega}.$$

The quantity τ is referred to as the *periodic time* of the motion. The number of oscillations that the body performs per unit time is called the *frequency ν* and is defined in terms of the periodic time as

$$\nu = \frac{1}{\tau}.$$

Example 3.1 Consider a light spring, fixed at one end, that is free to oscillate vertically in the y-direction. From experiment it was deduced that a mass of 1 kg attached to the free end of the spring will be displaced by $\frac{1}{30}$ m when a force of 3 N is applied. Find the angular frequency, frequency and period of the motion.

The mass is extended by $\frac{1}{15}$ m from its equilibrium position and released with initial speed 2 ms^{-1} at time $t = 0$. Find the position of the mass at later time t, amplitude and phase shift.

Solution To find the angular frequency, we need to use Hooke's law(3.1) to first determine the spring constant. Thus

$$k = -\frac{F(y)}{y} = \frac{-(-3)}{1/30} = 900 \text{ kgs}^{-2}.$$

So, the angular frequency is

$$\omega = \sqrt{\frac{k}{m}} = 30 \text{ rads}^{-1},$$

the period of the motion is

$$\tau = \frac{2\pi}{\omega} = \frac{\pi}{15} \text{ s}$$

and the frequency is

$$\nu = \frac{1}{\tau} = \frac{15}{\pi} \text{ s}^{-1}.$$

Using (3.7), we have

$$y(t) = A\cos(30t - \phi)$$

and

$$\dot{y}(t) = -30A\sin(30t - \phi).$$

Thus, from the initial conditions,

$$-\frac{1}{15} = A\cos\phi$$

and

$$2 = 30A\sin\phi.$$

Therefore, the phase shift is

$$\phi = \tan^{-1}(-1) = -\frac{\pi}{4}$$

and the amplitude is

$$A = \frac{\sqrt{2}}{15} \text{ m.}$$

Finally, the position of the mass at a later time t is

$$y(t) = \sqrt{\frac{2}{15}} \cos(30t + \pi/4). \qquad \square$$

3.3   Period of small oscillations

Consider a body of mass m attached to one end of a spring, which is performing small oscillations on a smooth horizontal table about an equilibrium point $x = x_0$ (see Figure 3.5).

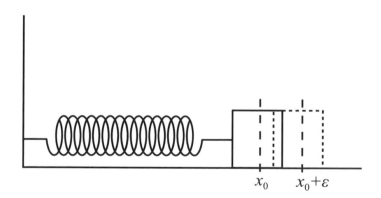

$x_0 \qquad x_0 + \varepsilon$

FIGURE 3.5
Body performing small oscillations about $x = x_0$

The force $F(x)$ displacing the body a small distance ϵ has potential energy $V(x)$ and is given by

$$F(x) = -\frac{dV}{dx}.$$

It was already established above that for this type of system the equilibrium position must be stable. This means that

$$\frac{dV(x_0)}{dx} = 0, \qquad \frac{d^2V(x_0)}{dx} > 0,$$

implying

$$F(x_0) = 0, \qquad \frac{dF(x_0)}{dx} < 0.$$

Now, for a small displacement ϵ, we have

$$x = x_0 + \epsilon.$$

Substituting this into the equation of motion (3.3) yields

$$m\ddot{\epsilon} = F(x_0 + \epsilon).$$

As ϵ is small (and remains small as the motion is stable), $F(x_0 + \epsilon)$ can be expanded using the Taylor series neglecting terms in ϵ^2 and higher. Thus

$$F(x_0 + \epsilon) = F(x_0) + \epsilon \frac{dF(x_0)}{dx},$$

giving

$$m\ddot{\epsilon} - \epsilon \frac{dF(x_0)}{dx} = 0.$$

This can be written as the simple harmonic motion equation

$$m\ddot{\epsilon} + \omega^2 \epsilon = 0,$$

where

$$\omega^2 = \frac{-1}{m} \frac{dF(x_0)}{dx}$$

or, equivalently,

$$\omega^2 = \frac{1}{m} \frac{d^2 V(x_0)}{dx^2}.$$

Example 3.2 A particle of mass m is attached to one end of a light elastic string of length 3 cm and the other end of the string is fixed at A, as shown in Figure 3.6. The tension in the string is denoted by f_T.

The particle is free to slide up and down a smooth vertical wire, where the minimum distance between the wire and A is 3 cm. The equilibrium position of the particle is at C, which is a distance 4 cm below B, and the angle between the string and the vertical wire at this point is α. Find the value of the spring constant k and the period of small oscillations of the particle about the equilibrium position.

Solution The equation of motion of the system is

$$m\ddot{y} = mg - f_T \cos \alpha.$$

The tension in the string f_T is, by Hooke's law,

$$f_T = ke,$$

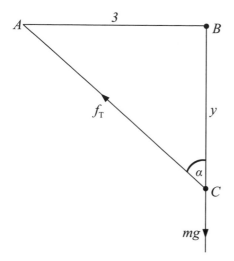

FIGURE 3.6
Period of small oscillations

where e is the extension of the string from its natural length. Thus

$$m\ddot{y} = mg - ke\cos\alpha$$

$$= mg - k\left(\sqrt{9+y^2} - 3\right)\left(\frac{y}{\sqrt{9+y^2}}\right)$$

$$= mg - ky\left(1 - \frac{3}{\sqrt{9+y^2}}\right).$$

When the particle is at equilibrium, $y = 4$ cm and $\ddot{y} = 0$. Substituting these into the above gives

$$k = \frac{5mg}{8}.$$

Therefore,

$$\ddot{y} = g\left[1 - \frac{5y}{8}\left(1 - \frac{3}{\sqrt{9+y^2}}\right)\right].$$

For a small displacement ϵ about the equilibrium position

$$y = 4 + \epsilon.$$

Substituting this into the above yields

$$\ddot{\epsilon} = g\left[1 - \frac{5}{8}(4+\epsilon)\left(1 - \frac{3}{\sqrt{25 + 8\epsilon + \epsilon^2}}\right)\right]$$

$$= g\left[1 - \frac{5}{8}\left(4+\epsilon - \frac{(12 + 3\epsilon)}{5}\left(1 + \frac{8\epsilon + \epsilon^2}{25}\right)^{-1/2}\right)\right]$$

$$= g\left[1 - \frac{5}{8}\left(4+\epsilon - \frac{(12 + 3\epsilon)}{5}\left(1 - \frac{4\epsilon}{25}\right)\right)\right],$$

where the last term in parentheses was obtained by performing a binomial expansion and excluding terms in ϵ of order ϵ^2 or higher. So,

$$\ddot{\epsilon} = g\left[1 - \frac{5}{8}\left(4+\epsilon - \frac{(12 + 3\epsilon)}{5} + \frac{48\epsilon}{125}\right)\right],$$

where, again, order of ϵ^2 has been excluded. The equation of simple harmonic motion can now be written as

$$\ddot{\epsilon} + \frac{49}{100}g\epsilon = 0$$

with $\omega^2 = \frac{49}{100}g$. Hence, the period of small oscillations τ_p is

$$\tau_p = \frac{2\pi}{\omega} = \frac{20\pi}{7\sqrt{g}}. \qquad \square$$

3.4 Damped simple harmonic motion

The model of simple harmonic motion given in previous sections is, of course, highly idealised. The body considered in Figure 3.1 was oscillating about its equilibrium position on top of a smooth table. The total energy remained constant and the oscillations would continue indefinitely. However, in practice, there will always be some dissipation of energy resulting from resistive forces which will *dampen* the simple harmonic motion.

For resistive forces that are not too large, it is reasonable to assume that they are proportional to the velocity as discussed in Chapter 2 (see Figure 3.7).

With this force incorporated, the equation of motion for the system in Figure 3.7 is

$$m\ddot{x} = -b\dot{x} - kx,$$

where b is the constant of proportionality. Rearranging the equation and dividing through by m yields the *damped simple harmonic motion equation*:

$$\ddot{x} + 2\kappa\dot{x} + \omega^2 x = 0, \qquad (3.9)$$

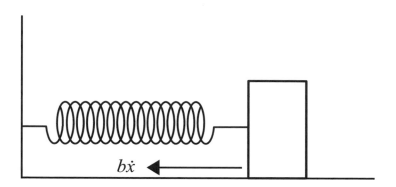

FIGURE 3.7
Body performing simple harmonic motion with a resistive or damping force $b\dot{x}$

where $2\kappa = b/m$ and $\omega^2 = k/m$. The particular form of κ is called the *damping constant* and will be considered in more detail presently. The equation (3.9) was chosen to facilitate the algebraic manipulation of the solutions, and again is a linear homogeneous second-order ordinary differential equation with constant coefficients. It is solved in the same manner as (3.4); the only difference is that there are three possible general solutions, each being determined for a different value of κ.

Assuming solutions $x(t) = e^{\lambda t}$ exist, the auxiliary equation corresponding to (3.9) is

$$\lambda^2 + 2\kappa\lambda + \omega^2 = 0.$$

Then the roots of this quadratic equation, in λ, are

$$\lambda = -\kappa \pm \sqrt{\kappa^2 - \omega^2}. \qquad (3.10)$$

One can see immediately that the discriminant can be negative, positive or zero depending on whether $\kappa < \omega$, $\kappa > \omega$ or $\kappa = \omega$, respectively. Each of these cases will be considered in turn.

Underdamped motion: $\kappa < \omega$
In this case, the discriminant in (3.10) is negative, giving λ a complex form

$$\lambda = -\kappa \pm i\Omega,$$

where $\Omega \equiv \sqrt{\omega^2 - \kappa^2}$. Thus the two linearly independent complex solutions are

$$x_1(t) = e^{(-\kappa + i\Omega)t}$$

and

$$x_2(t) = e^{(-\kappa - i\Omega)t}.$$

Following the procedure of Section 3.2, the real form of these solutions can be written as

$$x(t) = e^{-\kappa t}(B_1 \cos \Omega t + B_2 \sin \Omega t) \tag{3.11}$$

or, equivalently,

$$x(t) = Ae^{-\kappa t} \cos(\Omega t - \phi) \tag{3.12}$$

for real constants A, B_1, B_2 and ϕ.

The general solution (3.12) describes simple harmonic motion with an exponentially decaying amplitude $Ae^{-\kappa t}$, as depicted in Figure 3.8. The fre-

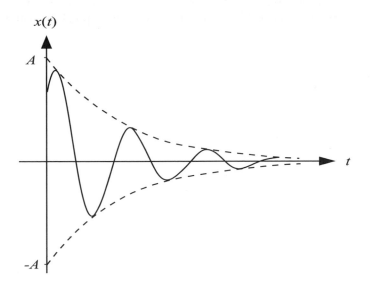

FIGURE 3.8
Underdamped motion

quency of the oscillations Ω will always be smaller than the frequency ω for undamped oscillations, and as in the undamped case, these oscillations will continue indefinitely.

Overdamped motion: $\kappa > \omega$
The roots of the auxiliary equation (3.10) are now real:

$$\lambda = -\kappa \pm \Omega,$$

where $\Omega \equiv \sqrt{\kappa^2 - \omega^2}$, and the two linearly independent solutions

$$x_1(t) = e^{(-\kappa + \Omega)t}$$

and

$$x_2(t) = e^{(-\kappa - \Omega)t}$$

are therefore also real. The general solution is a linear combination of these
two solutions

$$x(t) = C_1 e^{(-\kappa+\Omega)t} + C_2 e^{(-\kappa-\Omega)t}, \tag{3.13}$$

where C_1 and C_2 are arbitrary constants. It is clear that no oscillatory motion
can take place, and because $\kappa > \Omega$ both exponential terms will decay to zero
as $t \to \infty$. For example, if the system in Figure 3.2 were to be immersed in a
medium of sufficiently high viscosity, the body attached to the external spring
(when released) would take an infinite time to reach its equilibrium position.
In a slightly less viscous medium the body could overshoot and then approach
the equilibrium position from its other side, again taking an infinite time to
reach the equilibrium position. Figure 3.9 illustrates three possible instances

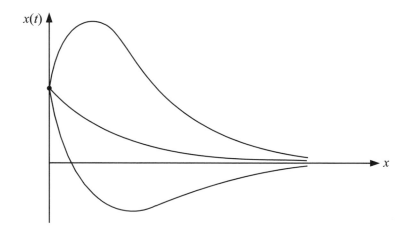

FIGURE 3.9
Overdamped motion

of such a motion.

Critically damped motion: $\kappa = \omega$
For this case, the auxiliary equation (3.10) still has two *repeated* roots

$$\lambda = -\kappa.$$

However, the solutions $x_1(t) = e^{-\kappa t}$ and $x_2(t) = e^{-\kappa t}$ are *not* linearly independent. They can be made so by multiplying one of the solutions by t. The
general solution is then the linear combination of these solutions:

$$x(t) = D_1 e^{-\kappa t} + D_2 t e^{-\kappa t}, \tag{3.14}$$

where D_1 and D_2 are arbitrary constants. One can easily show that this is a solution by substituting directly into (3.9). Notice that on comparing the arguments of the exponents in (3.13) and (3.14), we have

$$-\kappa - \Omega < -\kappa < -\kappa + \Omega.$$

This implies that as $t \to \infty$ the solution (3.14) decays to zero faster than (3.13) provided $C_1 \neq 0$. However, the displacement-time graph for (3.14) will be of the same form as Figure 3.9. Thus, for mechanisms in which it is desirable for the equilibrium position to be reached in the shortest possible time (galvinometer pointers, car shock absorbers, etc.) critical damping is the ideal situation.

3.5 Damped simple harmonic motion with a forcing term

What do a see-saw, Big Ben and the lights on the Eiffel Tower all have in common? They would not be able to function without a periodic *driving force*. For the see-saw, this is supplied by the kicks to the ground given alternately by each rider; Big Ben's pendulum oscillates due to its gravity escapement; and without an alternating current, the Eiffel Tower would be a mere shadow in the moonlight. Of course, to fully illuminate the Eiffel Tower's 5 billion bulbs, periodic injections of large quantities of cash are also requisite!

Denoting the driving force of an oscillator by $f(t)$ and noting that it is a function of time, it is placed on the right-hand side of the equation of motion with the other resistive forces. However, because $f(t)$ is a driving force and not a resistive one, it has a plus sign. Thence, the equation of motion with a forcing term is

$$m\ddot{x} = -b\dot{x} - kx + f(t)$$

or

$$m\ddot{x} + b\dot{x} + kx = f(t).$$

Rearranging this equation and dividing through by m yields

$$\ddot{x} + 2\kappa\dot{x} + \omega^2 x = F(t), \tag{3.15}$$

where $2\kappa = b/m$, $\omega^2 = k/m$ and $F(t) = f(t)/m$, which is the force per unit mass.

Equation (3.15) is a linear second-order *non-homogeneous* ordinary differential equation with constant coefficients. The technique used for determining solutions to this equation is the same as in the case of the homogeneous equation, except that we also require a function (any function) that satisfies the non-homogeneous equation. This function is called the *particular solution* or *particular integral* and to find it, we can make an educated guess. In cases where $F(t)$ is not too complicated the particular solution can be found fairly easily by inspection. Fortunately, the most important case from our perspective is the *temporal periodic driving force*

$$F(t) = F_0 \cos \omega_0 t, \qquad (3.16)$$

where F_0 is the amplitude per unit mass of the driving force and ω_0 is the angular frequency of the driving force, with both constants being positive.

Now, to find the particular solution $x_p(t)$, we will guess that it takes the form

$$x_p(t) = C_1 \cos \omega_0 t + C_2 \sin \omega_0 t, \qquad (3.17)$$

where C_1 and C_2 are constants to be found. This is not an unreasonable guess for the following reason: equation (3.16) contains a cosine term, so it is therefore natural to assume that the particular solution will also contain a cosine term. However, a solution with cosine alone will, when differentiated, yield a sine term and this would place an unnatural restriction on κ in (3.15) — it would force κ to be zero on equating cosine and sine terms in the equation. This problem can be circumvented by incorporating a sine term as part of the particular solution. Thence (3.17).

Substituting (3.16) and (3.17) into (3.15) yields

$$-C_1 \omega_0^2 \cos \omega_0 t - C_2 \omega_0^2 \sin \omega_0 t - 2\kappa C_1 \omega_0 \sin \omega_0 t$$
$$+ 2\kappa C_2 \omega_0 \cos \omega_0 t + C_1 \omega^2 \cos \omega_0 t + C_2 \omega^2 \sin \omega_0 t = F_0 \cos \omega_0 t.$$

Equating coefficients of cosine and sine gives

$$-C_1 \omega_0^2 + 2C_2 \kappa \omega_0 + C_1 \omega^2 = F_0$$

and

$$-C_2 \omega_0^2 - 2C_1 \kappa \omega_0 + C_2 \omega^2 = 0.$$

Solving for C_1 and C_2:

$$C_1 = \frac{(\omega^2 - \omega_0^2) F_0}{(\omega^2 - \omega_0^2)^2 + (2\kappa \omega_0)^2}$$

and

$$C_2 = \frac{2\kappa \omega F_0}{(\omega^2 - \omega_0^2)^2 + (2\kappa \omega_0)^2}.$$

A more conventional form of (3.17) is

$$x_p(t) = A_0 \cos(\omega_0 t - \phi_0), \qquad (3.18)$$

where $C_1 = A_0 \cos \phi_0$ and $C_2 = A_0 \sin \phi_0$. On squaring and adding, we find

$$A_0^2 = \frac{F_0^2}{(\omega^2 - \omega_0^2)^2 + (2\kappa \omega_0)^2}, \qquad (3.19)$$

$$\phi_0 = \tan^{-1}\left(\frac{2\kappa\omega_0}{\omega^2 - \omega_0^2}\right), \qquad 0 < \phi_0 < \pi, \tag{3.20}$$

where A_0 is the amplitude of the response to the driving force and ϕ_0 is the *phase shift*, which measures the lag between the motion of an oscillator and the driving force.

The complete general solution of (3.15) is a combination of the particular solution $x_p(t)$ and the *complementary function* $x_c(t)$. The complementary function is merely the solution of the homogeneous equation (3.15) i.e. $F(t) = 0$. This was already obtained in Section 3.4 for three different cases of damping. In all three cases the solution contained decaying exponential terms.

Let us then give the complementary function the generic form

$$x_c(t) = A_1 e^{\lambda_1 t} + A_2 e^{\lambda_2 t},$$

where λ_1 and λ_2 are the roots of the auxiliary equation corresponding to the homogeneous equation $F(t) = 0$, with A_1 and A_2 being arbitrary constants. Thence, the general solution of (3.15) is

$$\begin{aligned} x(t) &= x_p(t) + x_c(t) \\ &= A_0 \cos(\omega_0 t - \phi_0) + A_1 e^{\lambda_1 t} + A_2 e^{\lambda_2 t}. \end{aligned} \tag{3.21}$$

Recall that in all three previous cases of damping, the real part of λ was less than zero: $\mathrm{Re}\lambda < 0$. Notice, then, that as $t \to \infty$ the exponential terms in (3.21) approach zero. For this reason the complementary function is called the *transient*. Regardless of any given initial conditions the motion of the oscillator is eventually determined by the driving force. For this reason, the complementary function is often referred to as the *steady-state solution* or the *forced response*. Note that as the initial conditions must be satisfied, the transient term must be included.

3.5.1 Resonance

Many systems that have the potential to oscillate also have the potential to be destroyed by these oscillations under appropriate circumstances. For even the smallest of driving forces can induce quite large vibrations in a system that on occasion produces catastrophic consequences. A well-documented example is that of London's Millennium Bridge. The normal human gait induced lateral oscillations of the footbridge that were continuously reinforced as more pedestrians crossed. As the bridge swayed, so did the loading pedestrians — in rapport — increasing the amplitude of oscillations and thereby reinforcing the sway. A more devastating example is that of the original Tacoma Narrows Bridge in Washington USA, only this time the driving force was wind-induced.

The phenomenon that led to the behaviour of the above systems is called *resonance* and is characterised by an oscillator being driven at the appropriate frequency to produce a greatly exaggerated response to a driving force.

Consider the amplitude (3.19) for the steady-state solution (3.18) with driving force (3.16):

$$A_0 = \frac{F_0}{\sqrt{(\omega^2 - \omega_0^2)^2 + 4(\kappa\omega_0)^2}}. \tag{3.22}$$

The amplitude A_0 for forced oscillatory motion, keeping F_0 and κ fixed, can vary quite considerably if it is either a function of the driver frequency ω_0 or the undamped frequency ω of the oscillating system. Assume then that A_0 depends on ω_0 with ω fixed. As $\omega_0 \to 0$, $A_0 \to F_0/\omega^2$, and as $\omega_0 \to \infty$, $A_0 \to 0$. The maximum amplitude will occur when the denominator of (3.22) is a minimum. Thus, let

$$f(\omega_0^2) = (\omega^2 - \omega_0^2)^2 + 4(\kappa\omega_0)^2.$$

Differentiating with respect to ω_0^2 and equating to zero yields

$$\omega_0^2 = \omega^2 - 2\kappa^2 = \omega^2\left(1 - 2\frac{\kappa^2}{\omega^2}\right), \tag{3.23}$$

where ω_0 is known as the *resonant frequency* in this case. Notice that the maximum amplitude will occur at a frequency that is a little less than the undamped frequency:

$$A_{\max} = \frac{F_0}{2\kappa\sqrt{\omega^2 - \kappa^2}}.$$

If $2\kappa^2/\omega^2 > 1$ then ω_0 will be imaginary. So the maximum amplitude will occur for $\omega_0 = 0$ and $A_{\max}$ will be a monotonic decreasing function of ω_0.

For small values of the damping constant κ, we can from (3.23) make the approximation $\omega_0 \approx \omega$. Thence, from (3.22)

$$A_{\max} \approx \frac{F_0}{2\kappa\omega},$$

which indicates that larger maximum amplitudes occur for small values of the damping constant. This is illustrated in Figure 3.10. The phase shift (3.20) indicates by how much the oscillatory motion lags behind the driving force. In particular, if ϕ varies with respect to ω_0 and κ is small, the oscillator's motion and the driving force will almost be in phase if $\omega_0 \ll \omega$. For $\omega_0 = \omega$, $\phi = \pi/2$, so the oscillator's motion lags behind the driving force by $\pi/2$.

Example 3.3 A vertical spring with spring constant 5 Nm^{-1} is fixed at one end while a body of mass 1 kg is allowed to hang freely at the other end. The body is then extended 2 m in the vertical direction and released. The body is subject to a vertical driving force of $\alpha \cos\omega t$ (α is a positive constant) and a resistive force $2\dot{x}$. Obtain the general solution of the motion, the amplitude of the steady-state solution, and the phase shift. Furthermore, find the maximum amplitude and determine the angular frequency for which this occurs.

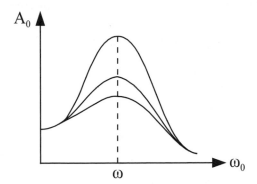

FIGURE 3.10
Amplitude against angular frequency for a driven oscillator. The curves in descending order: $\kappa/\omega = 0.1, 0.2, 0.3$

Solution The equation of motion and initial conditions are

$$\ddot{x} + 2\dot{x} + 5x = \alpha \cos \omega t$$

$$x(0) = 2, \quad \dot{x}(0) = 0.$$

The complementary function $x_c(t)$ is given for

$$\ddot{x} + 2\dot{x} + 5x = 0.$$

The auxiliary equation is then

$$\lambda^2 + 2\lambda + 5 = 0$$

yielding roots

$$\lambda = -1 \pm 2i.$$

Therefore,

$$x_c(t) = (C_1 \cos 2t + C_2 \sin 2t)e^{-t},$$

where C_1 and C_2 are arbitrary constants to be found for the given initial conditions.

The right side of the equation of motion suggests that an appropriate trial solution would be

$$x_p(t) = D_1 \cos \omega t + D_2 \sin \omega t$$

for constants D_1 and D_2 to be obtained. Substituting this into the equation of motion and comparing coefficients of sine and cosine yields

$$D_1 = \frac{\alpha(5 - \omega^2)}{(5 - \omega^2)^2 + (2\omega)^2}, \quad D_2 = \frac{2\alpha\omega}{(5 - \omega^2) + (2\omega)^2}.$$

The general solution to the equation of motion can then be written as

$$x(t) = A_0 \cos(\omega t - \phi) + (C_1 \cos 2t + C_2 \sin 2t)e^{-t},$$

where the amplitude of the steady-state solution is

$$A_0 = \frac{\alpha}{\sqrt{(5 - \omega^2)^2 + (2\omega)^2}}$$

and the phase shift is

$$\phi = \tan^{-1}\left(\frac{2\omega}{5 - \omega^2}\right).$$

The initial conditions must now be incorporated to obtain values for C_1 and C_2. The first initial condition $x(0) = 2$ gives

$$2 = A_0 \cos \phi + C_1$$
$$\Rightarrow C_1 = 2 - A_0 \cos \phi.$$

The second initial condition $\dot{x}(0) = 0$ gives

$$0 = A_0 \omega \sin \phi + 2C_2 - C_1$$
$$\Rightarrow C_2 = 1 - (A_0/2) \cos \phi - (A_0 \omega/2) \sin \phi.$$

Now, as α is a constant, the amplitude is a maximum when the function

$$f(\omega^2) = (5 - \omega^2)^2 + (2\omega)^2$$

is a minimum. Differentiating this with respect to ω^2 gives

$$f'(\omega^2) = -2(5 - \omega^2) + 4 = 0 \qquad \therefore \omega = \sqrt{3}.$$

Thus $\omega = \sqrt{3}$ rads^{-1} is the necessary angular frequency for a maximum amplitude

$$A_{\max} = \frac{\alpha}{4}.$$

Note that if some reasonable values of α and ω were substituted into the solution of Example 3.3, and a graph plotted of $x(t)$ against t, one would see that the transient term would quickly fade away (in less than one complete cycle) leaving just the steady-state solution. (Try this for some values of α and ω of your choice.)   □

3.6 The LCR circuit

It is of some interest to consider briefly a different but analogous system to that of the previous sections. It is called the LCR circuit and comprises an *inductor*(L), a *capacitor*(C), and a *resistor*(R) placed in series as shown in Figure 3.11.

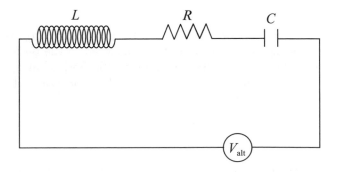

FIGURE 3.11
A simple LCR circuit

The *voltage* V_{alt} applied to the circuit induces an *alternating current* I to flow through it. The equation of motion of the LCR circuit is then

$$L\ddot{q} + R\dot{q} + \frac{1}{C}q = V_{\text{alt}}(t), \quad C \neq 0,$$

where $q(t)$ is the charge on the capacitor and is related to the current by $\dot{q}(t) = I$. Notice that L, $R\dot{q}$ and $1/C$ are analogous to mass, resistive force and spring constant, respectively, for a mechanical oscillatory system. Then V_{alt} would correspond to the driving force. If $V_{\text{alt}} = 0$, the electrical oscillatory motion is said to be *free*. The auxiliary equation is given by

$$L\lambda^2 + R\lambda + \frac{1}{C} = 0$$

yielding the roots

$$\lambda = \frac{-R \pm \sqrt{R^2 - 4L/C}}{2L}.$$

Provided $R \neq 0$, we can define the following three alternatives for the circuit:

$$\text{underdamped}: R^2 - \frac{4L}{C} < 0$$

$$\text{overdamped}: R^2 - \frac{4L}{C} > 0$$

$$\text{critically damped}: R^2 - 4LC = 0.$$

As was the case for the mechanical system, this electrical system will yield a general solution that contains an exponential term for all the above three alternatives: namely, $e^{-Rt/2L}$. This means as $t \to \infty$, $q(t) \to 0$, and only the steady-state solution (particular solution) remains.

3.7   Exercises

1. A body sits on a smooth horizontal table. A light elastic cord of natural length a is fixed at the origin O and attached to the body at the other end. The body is free to move in the horizontal x-direction. The body is then extended by a length a and released. Using Hooke's law in the form

$$f_T = \frac{\lambda e}{a},$$

where λ is the constant *modulus of elasticity*, e is the extension and a the natural length of a cord, show that the body will pass the origin O at a time

$$t = \frac{1}{2\omega}(2 + \pi), \quad \omega^2 = \frac{\lambda}{ma}$$

after release. (Gravity may be neglected.)

2. A body of mass m is suspended from one end of a light spring that hangs vertically and the other end of the spring is fixed. The natural length of the spring is y_0 and it has length y_1 at equilibrium. If the body is then extended a distance $3y_1$ and released, find the frequency of the subsequent oscillations and show that the body is displaced from its equilibrium position by an amount $3y_1 \cos \omega t$.

3. Find the period of small oscillations of the particles in Exercises 2.3 and 2.4, about their equilibrium positions.

4. The equation of motion for a given damped oscillator is

$$m\ddot{x} + b\dot{x} + kx = 0$$

with initial conditions

$$x(0) = x_0, \quad \dot{x}(0) = v_0.$$

Assuming that the motion is underdamped, find the general solution. Obtain the amplitude and phase shift of the motion. What happens to the amplitude as $b \to \sqrt{4mk}$?

5. A particle of mass m is restricted to move along the positive x-axis. It is subject to a force with potential energy

$$V = \frac{m\omega^2}{2x^2}(x^4 + a^4),$$

where ω and a are constants. Draw the potential energy diagram. Determine the equilibrium position and obtain the period of small oscillations about this position.

6. A particle of mass m is restricted to move along the positive x-axis. It is subject to a force with potential energy

$$V = V_0 \left(\frac{x}{x_0} + \gamma^2 \frac{x_0}{x} \right),$$

where V_0, x_0 and γ are positive constants. Draw the potential energy diagram. Determine the equilibrium position and obtain the period of small oscillations about this position.

7. Two springs S_1 and S_2 hang parallel to each other and are fixed at their top end. A force of 50 N stretches S_1 by 5 m and a force of 160 N stretches S_2 by 2 m. Masses of 10 kg and 20 kg are attached to the free ends of S_1 and S_2, respectively. The mass of 10 kg and the mass of 20 kg are then released from their equilibrium positions with upwards velocities of 1 ms^{-1} and 2 ms^{-2}, respectively. Show that the amplitudes of S_1 and S_2 are the same. What are the velocities of each mass at $t = \pi$? Find the times for which the displacements of the masses from their equilibrium positions are the same. In which directions are they moving at these times?

8. The equation of motion for a given driven oscillator is

$$\ddot{x} + \omega^2 x = F(t)$$

with initial conditions

$$x(0) = 0, \quad \dot{x}(0) = 0.$$

Show that a solution to this equation for $F(t) = F_0 \cos \omega_0 t$ is

$$x(t) = \frac{F_0}{\omega^2 - \omega_0^2}(\cos \omega_0 t - \cos \omega t). \tag{3.24}$$

Use L'Hospital's rule[3] to show that

$$\lim_{\omega_0 \to \omega} x(t) = \frac{F_0 t}{2\omega} \sin \omega t.$$

---

[3]For our purposes it will suffice that for two functions $f(x)$ and $g(x)$, $\lim_{x \to a} f(x)/g(x)$ is equal to $\lim_{x \to a} f'(x)/g'(x)$ if $f(a) \to 0$ and $g(a) \to 0$.

Obtain the same result if $F(t) = F_0 \cos \omega t$. Show further that

$$x(t) = \frac{-2F_0}{\omega^2 - \omega_0^2} \sin \frac{1}{2}(\omega_0 - \omega)t \, \sin \frac{1}{2}(\omega_0 + \omega)t$$

is equivalent to (3.24) and that for $\delta \ll 1$

$$x(t) \approx \frac{F_0}{2\delta\omega_0} \sin \delta t \sin \omega_0 t$$

for an appropriate choice of δ.

9. A particle P of mass m is restricted to move along the x-axis under the action of a force $m\omega^2 x$ toward an origin O, where ω is a positive constant and $OP = x$. The motion is resisted by a force $2am\omega v$, where v is the speed and $0 < a < 1$. Find the solution to the equation of motion given the initial conditions

$$x(0) = 0 \quad \text{and} \quad \dot{x}(0) = u.$$

Let the ratio of the distances from O of consecutive positions of rest be γ. Show that

$$\frac{a^2\pi^2}{1 - a^2} = (\ln \gamma)^2.$$

10. A mass of 2 kg is attached to a vertical spring with the other end of the spring fixed. The spring constant is 4 Nm^{-1}. The mass is subject to a vertical driving force of $8 \cos \omega t$ and a resistive force of $8\dot{x}$. Assuming that the mass is initially at rest at its equilibrium position, find the solution and amplitude A of the steady-state part in terms of ω. Draw a graph of A against ω and determine the maximum amplitude and the angular frequency for which this occurs.

11. Given the initial conditions $q(0) = q_0$ and $I(0) = 0$ find the charge $q(t)$ on a capacitor in an LCR circuit when the components have the following values: $L = 1/2$ henry, $C = 1/125$ farad, $R = 5$ ohms and $V_{\text{alt}} = 0$ volts.

12. The equation of motion of an LCR circuit with a driving force of $V_{\text{alt}} = V_0 \sin \omega t$ is given by

$$L\ddot{q}(t) + R\dot{q}(t) + C^{-1}q = V_0 \sin \omega t.$$

Show that the steady-state solution is given by

$$q(t) = \frac{-V_0}{\omega Z^2} \sin(\omega t + \tan^{-1}(R/X)),$$

where $X = (LC - 1)\omega/C$ (ohms) is known as the *reactance* and $Z = (X^2 + R^2)^{1/2}$ is known as the *impedance* (ohms).

Also, show that the steady-state current $I(t) = \dot{q}(t)$ is given by

$$I(t) = \frac{V_0}{Z^2}\sin(\omega t - \tan^{-1}(X/R)).$$

What is the amplitude of the steady-state current?

4

Particle Dynamics in Two and Three Dimensions

CONTENTS

Modelling the dynamical behaviour of particles in one dimension can be very illuminating but restrictive. If one wishes to discuss the motion of a projectile, particles travelling through an electromagnetic field or, as we will encounter in Chapter 5, the orbital motion of particles then a two- or three-dimensional approach becomes necessary.

4.1 Projectiles

The term *projectile* is usually restricted to particles and bodies that travel through space in a vertical plane under only the action of gravity and perhaps air resistance. Bodies in flight that are aided by an internal propulsion system, such as rockets and aeroplanes, are not considered to be projectiles.

When considering the motion of a projectile in a vertical plane, we will ignore any effects due to the Earth's rotation; the gravitational field will be assumed uniform and resistive forces will eventually be incorporated.

4.1.1   Projectiles without air resistance

The equation of motion of a projectile subject only to gravity close to the
Earth's surface is given by

$$m\ddot{\mathbf{r}} = m\mathbf{g}, \tag{4.1}$$

where $\mathbf{g} = (0, -g, 0)$ is the gravitational acceleration of constant magnitude
$|\mathbf{g}| \equiv g \approx 10$ ms^{-2}. If the projectile is assumed to move only in the xy-plane,
then (4.1) can be written as

$$m\ddot{\mathbf{r}} = -mg\mathbf{j}, \tag{4.2}$$

where $\mathbf{r} = \mathbf{r}(x, y, 0)$ is the position vector of the projectile at any time t and
$\mathbf{j}$ is the unit normal vector in the y-direction, as indicated in Figure 4.1.

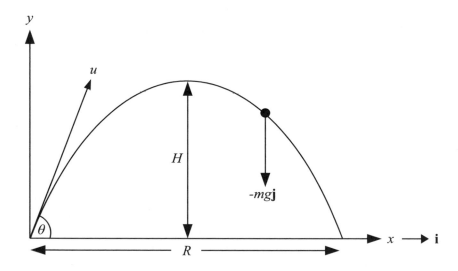

FIGURE 4.1
Path of a projectile

The projectile is being launched with *initial speed* u at an angle θ to the
horizontal. The velocity of the projectile at any point along its trajectory is
a linear combination of the speeds in the x- and y-directions. Thus, initially
$(t = 0)$, $\mathbf{r} = \mathbf{0}$ and $\dot{\mathbf{r}} = (\dot{x}, \dot{y}, \dot{z}) = (u\cos\theta, u\sin\theta, 0)$. (Note that we are
adopting the abbreviated forms $\dot{\mathbf{r}} = d\mathbf{r}/dt$ and $\mathbf{r} = (x, y, z) \equiv x\mathbf{i} + y\mathbf{j} + z\mathbf{k}$.)
The scalar form of the equations of motion are, from (4.2),

$$\ddot{x} = 0, \quad \ddot{y} = -g, \quad \ddot{z} = 0.$$

On integrating and incorporating the initial conditions, we have

$$\dot{x} = u\cos\theta, \quad \dot{y} = u\sin\theta - gt, \quad \dot{z} = 0.$$

One more integration then yields the horizontal and vertical displacements of the projectile:

$$x = ut \cos \theta \tag{4.3}$$

$$y = ut \sin \theta - \frac{1}{2} g t^2 \tag{4.4}$$

$$z = 0.$$

When $y = 0$, we have the following quadratic equation in t:

$$t\left(u \sin \theta - \frac{1}{2} g t \right) = 0,$$

with two roots $t = 0$ and $t = 2(u/g) \sin \theta$. The second set gives the *time of flight* of the projectile; that is, the time which the projectile takes to cover the distance R in Figure 4.1. Substituting the time of flight equation into (4.3) yields

$$R = \frac{u^2}{g} \sin 2\theta, \tag{4.5}$$

where R is the *horizontal range* of the projectile. The *maximum range* is attained by putting $\sin 2\theta = 1$, which is equivalent to $\theta = \pi/4$. This then gives

$$R_{\max} = \frac{u^2}{g}. \tag{4.6}$$

That the locus of the projectile's path in the xy-plane is a concave downward parabola can be most easily verified by eliminating t in (4.3) and (4.4). Thus,

$$y = x \tan \theta - \frac{g}{2u^2} x^2 \sec^2 \theta. \tag{4.7}$$

It can clearly be seen from Figure 4.1 that the projectile will attain its maximum height H when $dy/dx = 0$. Thus, from (4.7)

$$\frac{dy}{dx} = \tan \theta - \frac{g}{u^2} x \sec^2 \theta = 0,$$

giving

$$x = \frac{u^2}{2g} \sin 2\theta. \tag{4.8}$$

Notice that this is half the range on comparing with (4.5), which is not unexpected due to the symmetry of the projectile's trajectory. Substituting (4.8) into (4.7) gives

$$H = \frac{u^2}{2g} \sin^2 \theta. \tag{4.9}$$

Example 4.1 A projectile is launched with speed u at an angle θ_1 to a plane that is itself inclined at an angle θ_2 to the horizontal. If

θ_2 is constant, find the time of flight and the maximum range of the projectile. (Ignore air resistance.)

Solution It will be convenient to choose axes parallel and perpendicular to the plane (see Figure 4.2).

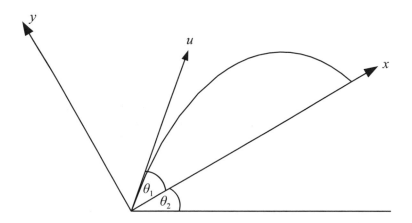

FIGURE 4.2
Path of a projectile on an inclined plane

The scalar form of the equations of motion are

$$\ddot{x} = -g \sin \theta_2$$
$$\ddot{y} = -g \cos \theta_2.$$

Assuming the initial conditions $x(0) = y(0) = 0$ and $\dot{x}(0) = u \cos \theta_1$, $\dot{y}(0) = u \sin \theta_1$, we have on twice integrating the equations of motion

$$x = ut \cos \theta_1 - \frac{1}{2} g t^2 \sin \theta_2$$
$$y = ut \sin \theta_1 - \frac{1}{2} g t^2 \cos \theta_2. \qquad (4.10)$$

The time of flight is found when $y = 0$. Thus, ignoring $t = 0$, we have

$$t = \frac{2u}{g} \sin \theta_1 \sec \theta_2.$$

The range of the projectile is found by substituting the time of flight into (4.10), giving

$$R = \frac{2u^2}{g} \sin\theta_1 \sec^2\theta_2 (\cos\theta_1 \cos\theta_2 - \sin\theta_1 \sin\theta_2)$$

$$= \frac{2u^2}{g} \sin\theta_1 \sec^2\theta_2 \cos(\theta_1 + \theta_2)$$

$$= \frac{u^2}{g} \sec^2\theta_2 [\sin(2\theta_1 + \theta_2) - \sin\theta_2].$$

The maximum range of the projectile occurs when $\sin(2\theta_1 + \theta_2) = 1$; that is, $2\theta_1 + \theta_2 = \pi/2$. As θ_2 is constant, we have $\theta_1 = \pi/4 - \theta_2/2$, yielding

$$R_{\max} = \frac{u^2}{g}(1 + \sin\theta_2)^{-1}. \qquad \square$$

4.1.2 Projectiles with linear air resistance

The equation of motion with linear air resistance $-k\dot{\mathbf{r}}$ is given by

$$m\ddot{\mathbf{r}} = m\mathbf{g} - k\dot{\mathbf{r}}, \tag{4.11}$$

where k is a positive constant. Solving this vector equation is straightforward, and the scalar equations of position can then be retrieved.

The equation of motion (4.11) is a *first*-order linear differential equation $m\dot{\mathbf{r}}$ with constant coefficients. To solve for $\dot{\mathbf{r}}$ first, we can write this equation in the form

$$\frac{d}{dt}\dot{\mathbf{r}} + \frac{k}{m}\dot{\mathbf{r}} = \mathbf{g}$$

and choose an integrating factor $e^{kt/m}$. Multiplying through by the integrating factor yields

$$\frac{d}{dt}(e^{kt/m}\dot{\mathbf{r}}) = e^{kt/m}\mathbf{g}.$$

On integrating, we have

$$\dot{\mathbf{r}} = \frac{m}{k}\mathbf{g} + \mathbf{A}e^{-kt/m},$$

where $\mathbf{A}$ is a constant vector of integration. Let us assume that the projectile is launched at $t = 0$ with velocity $\mathbf{u}$, then $\mathbf{A} = \mathbf{u} - m\mathbf{g}/k$ and

$$\dot{\mathbf{r}} = \frac{m}{k}\mathbf{g} + \left(\mathbf{u} - \frac{m}{k}\mathbf{g}\right)e^{-kt/m}.$$

A second integration gives

$$\mathbf{r}(t) = \frac{m}{k}\mathbf{g}t - \frac{m}{k}\left(\mathbf{u} - \frac{m}{k}\mathbf{g}\right)e^{-kt/m} + \mathbf{B},$$

where $\mathbf{B}$ is a constant vector of integration. Assuming the initial conditions $\mathbf{r}(0) = \mathbf{0}$, we find that $\mathbf{B} = (m/k)(\mathbf{u} - m\mathbf{g}/k)$, giving the position vector of the projectile at any time t as

$$\mathbf{r}(t) = \frac{m}{k}\mathbf{g}t + \frac{m}{k}\left(\mathbf{u} - \frac{m}{k}\mathbf{g}\right)(1 - e^{-kt/m}). \qquad (4.12)$$

With reference to Figure 4.1, we can write down the scalar equations for the projectile's position. This is achieved by comparing components of $\mathbf{i}, \mathbf{j}$ and $\mathbf{k}$ in (4.12):

$$x(t) = \frac{m}{k}(1 - e^{-kt/m})u\cos\theta$$

$$y(t) = \frac{mg}{k}t + \frac{m}{k}\left(\frac{mg}{k} + u\sin\theta\right)(1 - e^{-kt/m})$$

$$z(t) = 0.$$

Notice that as $t \to \infty$, $x \to (m/k)u\cos\theta$ and $y \to -\infty$. Thus the projectile's trajectory is asymptotic to the vertical line $(m/k)u\cos\theta$, which also implies that the projectile falls almost vertically at the end of its trajectory.

4.2   Energy and force

In Section 2.1, we defined two important quantities for a particle undergoing rectilinear motion and subject to a force dependent on the particle's position; namely, the kinetic and potential energy of the particle. This then led to the conservation of energy law (2.4). As these concepts were developed for one-dimensional motion, let us consider how these notions translate in $\mathbb{E}^3$.

4.2.1   Work and kinetic energy

The equation of motion of a particle P with constant mass m under the action of a force $\mathbf{F} = \mathbf{F}(\mathbf{r})$ is given vectorially as

$$m\ddot{\mathbf{r}} = \mathbf{F}, \qquad (4.13)$$

which can be written, on taking the scalar product with $\dot{\mathbf{r}}$, as

$$m\dot{\mathbf{r}} \cdot \ddot{\mathbf{r}} = \mathbf{F} \cdot \dot{\mathbf{r}}. \qquad (4.14)$$

Notice that

$$\frac{1}{2}m\frac{d}{dt}(\dot{\mathbf{r}} \cdot \dot{\mathbf{r}}) = \frac{1}{2}m(\ddot{\mathbf{r}} \cdot \dot{\mathbf{r}} + \dot{\mathbf{r}} \cdot \ddot{\mathbf{r}}) = m\dot{\mathbf{r}} \cdot \ddot{\mathbf{r}}.$$

Thus (4.14) becomes

$$\frac{1}{2}m\frac{d}{dt}(\dot{\mathbf{r}} \cdot \dot{\mathbf{r}}) = \mathbf{F} \cdot \dot{\mathbf{r}}$$

or, equivalently,

$$\frac{d}{dt}\left(\frac{1}{2}m|\dot{\mathbf{r}}|^2\right) = \mathbf{F}\cdot\dot{\mathbf{r}}.$$

By definition

$$T = \frac{1}{2}m|\dot{\mathbf{r}}|^2$$

is called the kinetic energy, which is analogous to the one-dimensional form given by $\frac{1}{2}m\dot{x}^2$ (see (2.4)). Thus, the rate of change of kinetic energy of a particle P acted on by a force $\mathbf{F}$ is

$$\dot{T} \equiv \frac{dT}{dt} = \mathbf{F}\cdot\dot{\mathbf{r}}. \tag{4.15}$$

Now, consider the particle P under the action of a force $\mathbf{F}$, traversing a path $\mathcal{C}$ that lies between two points with position vectors $\mathbf{r}_1 = \mathbf{r}(t_1)$ and $\mathbf{r}_2 = \mathbf{r}(t_2)$, respectively (see Figure 4.3).

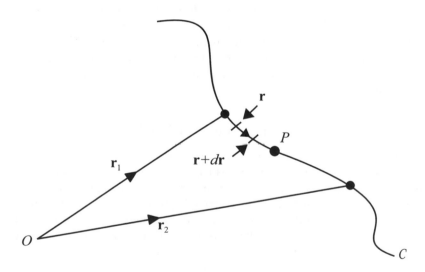

FIGURE 4.3
Particle traversing between two points under the action of $\mathbf{F}$

Integrating (4.15) over the time interval $[t_1, t_2]$ yields

$$T(t_2) - T(t_1) = \int_{t_1}^{t_2} \mathbf{F}(\mathbf{r})\cdot\dot{\mathbf{r}}dt$$

$$= \int_{t_1}^{t_2} \mathbf{F}(\mathbf{r})\cdot\frac{d\mathbf{r}}{dt}dt$$

$$= \int_{\mathcal{C}} \mathbf{F}(\mathbf{r})\cdot d\mathbf{r}. \tag{4.16}$$

As in the one-dimensional case, $\mathbf{F}(\mathbf{r}) \cdot d\mathbf{r}$ is the work done dW by the force $\mathbf{F}(\mathbf{r})$ in moving P from $\mathbf{r}$ to $\mathbf{r} + d\mathbf{r}$. Thence,

$$W_{12} = \int_C \mathbf{F}(\mathbf{r}) \cdot d\mathbf{r}, \tag{4.17}$$

where W_{12} is the work done under $\mathbf{F}$ in moving P from the point with position vector $\mathbf{r}_1$ to the point with position vector $\mathbf{r}_2$ along the path C. Again, as in the one-dimensional case, the quantity under the integral in (4.17) represents the *power*. Then

$$\text{Power} \equiv \frac{dW}{dt} = \mathbf{F} \cdot \mathbf{r},$$

which is also the rate of kinetic energy:

$$\text{Power} = \dot{T}.$$

The integral in (4.17) is an example of a *line integral*. It is a generalisation of the *ordinary* integrals exhibited in Section 2.1. In general, the line integral depends on the *path* of the particle in going between two points in $\mathbb{E}^3$, which is subtly different to the one-dimensional interval between two points along the x-axis, say. Nevertheless, the line integral reduces to the ordinary integral in the case of one-dimensional motion.

4.2.2 Potential energy and conservative force

The conservation of energy theorem depends on the fact that the total energy remains constant as the particle P moves from position $\mathbf{r}_1$ to position $\mathbf{r}_2$. The type of force that must act on P to achieve this result is called a *conservative force* and it must be a function of the particle's position alone. That is, forces that may depend on other quantities such as frictional forces, for example, are not conservative forces, and the conservation of energy theorem would not apply in these cases. Furthermore, the work done in moving the particle from $\mathbf{r}_1$ to $\mathbf{r}_2$ must necessarily be *path independent*. If this were not the case, the work done in moving the particle between two points could have different values for *any* path C.

If $\mathbf{F}(\mathbf{r})$ is a conservative force field, then by definition

$$\mathbf{F}(\mathbf{r}) = -\boldsymbol{\nabla} V(\mathbf{r}), \tag{4.18}$$

where $V(\mathbf{r})$ is the *scalar* potential energy (the negative sign is for later convenience). The *vector* function $\boldsymbol{\nabla}$ is a vector differential operator[1] and is defined as

$$\boldsymbol{\nabla} = \frac{\partial}{\partial x}\mathbf{i} + \frac{\partial}{\partial y}\mathbf{j} + \frac{\partial}{\partial z}\mathbf{k} \equiv \left(\frac{\partial}{\partial x}, \frac{\partial}{\partial y}, \frac{\partial}{\partial z} \right).$$

---

[1] The symbol $\boldsymbol{\nabla}$ has a number of different names: 'del', 'nabla', and 'grad.'

Note that (4.18) is the three-dimensional analogue of (2.2). Substituting (4.18) into (4.16) yields

$$T(t_2) - T(t_1) = \int_{\mathcal{C}} \mathbf{F}(\mathbf{r}) \cdot d\mathbf{r} = -\int_{\mathcal{C}} \nabla V(\mathbf{r}) \cdot d\mathbf{r}$$

$$= -\int_{\mathcal{C}} \left(\frac{\partial V}{\partial x}, \frac{\partial V}{\partial y}, \frac{\partial V}{\partial z} \right) \cdot (dx, dy, dz)$$

$$= -\int_{\mathcal{C}} \left(\frac{\partial V}{\partial x} dx + \frac{\partial V}{\partial y} dy + \frac{\partial V}{\partial z} dz \right)$$

$$= -\int_{\mathcal{C}} dV$$

$$= -(V(\mathbf{r}(t_2)) - V(\mathbf{r}(t_1))). \tag{4.19}$$

On rearranging, we have

$$T(t_2) + V(\mathbf{r}(t_2)) = T(t_1) + V(\mathbf{r}(t_1)),$$

which is the conservation of energy law:

$$T + V(\mathbf{r}) = E.$$

Note that the penultimate equality in (4.19) arises from the chain rule for the function $V = V(x, y, z)$.

> **Example 4.2** Find the work done under the force $\mathbf{F}(\mathbf{r}(t)) = (2xy, 3y, xz)$ in moving a particle along the path $\mathcal{C} : \mathbf{r}(t) = (t, t, t^2)$, where t is a parameter in the range $[0, 1]$. Is $\mathbf{F}$ a conservative force?
>
> **Solution** We have that $x = t$, $y = t$ and $z = t^2$. Then
>
> $$\dot{\mathbf{r}}(t) \equiv \frac{d\mathbf{r}(t)}{dt} = (1, 1, 2t)$$
>
> and
>
> $$\mathbf{F}(\mathbf{r}(t)) = (2t^2, 3t, t^3).$$
>
> Substituting these into (4.17) yields
>
> $$W = \int_{\mathcal{C}} \mathbf{F} \cdot d\mathbf{r} = \int_0^1 \mathbf{F}(\mathbf{r}(t)) \cdot \dot{\mathbf{r}}(t) dt$$
>
> $$= \int_0^1 (2t^2 + 3t + 2t^4) dt$$
>
> $$= \frac{2}{3} + \frac{3}{2} + \frac{2}{15}$$
>
> $$= \frac{69}{30}.$$

If **F** is conservative, there should exist a V such that

$$\mathbf{F} = -\nabla V \equiv -\left(\frac{\partial V}{\partial x}, \frac{\partial V}{\partial y}, \frac{\partial V}{\partial z}\right).$$

On comparing the components of $\mathbf{i}, \mathbf{j}$ and $\mathbf{k}$, we find that

$$\frac{\partial V}{\partial x} = -2xy, \quad \frac{\partial V}{\partial y} = -3y, \quad \frac{\partial V}{\partial z} = -xz.$$

But second-order partial derivatives must commute in their independent variables:

$$\frac{\partial^2 V}{\partial x \partial y} = \frac{\partial^2 V}{\partial y \partial x}, \quad \frac{\partial^2 V}{\partial x \partial z} = \frac{\partial^2 V}{\partial z \partial x}, \quad \frac{\partial^2 V}{\partial y \partial z} = \frac{\partial^2 V}{\partial z \partial y}.$$

As this condition is clearly violated in our case (except at zero), we conclude that there exists no V satisfying $\mathbf{F} = -\nabla V$ and consequently, $\mathbf{F}$ is not a conservative force. $\qquad\square$

Example 4.3 Consider a particle moving under a conservative force $\mathbf{F}(\mathbf{r}(t)) = (2ze^{2x}, z^2, 2yz + e^{2x})$ along a path $\mathcal{C} : \mathbf{r}(t) = (t, t, 1)$, where t is a parameter in the range $0 \le t \le 1$. Find the potential energy and the work done by the force in moving the particle along the path.

Solution $\mathbf{F}$ is conservative implies that $\mathbf{F} = -\nabla V$. On comparing components of $\mathbf{i}, \mathbf{j}$ and $\mathbf{k}$, we find that

$$\frac{\partial V}{\partial x} = -2ze^{2x} \tag{4.20}$$

$$\frac{\partial V}{\partial y} = -z^2 \tag{4.21}$$

$$\frac{\partial V}{\partial z} = -2yz - e^{2x}. \tag{4.22}$$

Now, even though $V = V(x, y, z)$, (4.20), (4.21) and (4.22) can be integrated straightforwardly, where the 'constant' of integration becomes a *function* of integration dependent on the remaining variables. Thus integrating (4.20) with respect to x yields

$$V(x, y, z) = \int -2ze^{2x} dx = -ze^{2x} + f(y, z), \tag{4.23}$$

where $f(y, z)$ is a function of integration. Differentiating this expression with respect to y and comparing with (4.21) gives

$$\frac{\partial V}{\partial y} = \frac{\partial f}{\partial y}(y, z) = -z^2,$$

implying that

$$f(y, z) = \int -z^2 dy = -z^2 y + g(z),$$ (4.24)

where $g(z)$ is a function of integration. To determine the potential energy function in (4.23), we are required to find $f(y, z)$ and hence, $g(z)$. This is achieved by differentiating (4.23) with respect to z and comparing with (4.22):

$$\frac{\partial V}{\partial z} = \frac{\partial f}{\partial z}(y, z) - e^{2x} = -2yz - e^{2x}$$

implying, on cancelling e^{2x}, that

$$f(y, z) = \int -2yz dz = -z^2 y + h(y),$$ (4.25)

where $h(y)$ is a function of integration. On comparing (4.24) and (4.25), we have

$$g(z) = h(y).$$

This is possible only if they were equal to the same *constant*. On grouping the various functions together, the potential energy in (4.23) is finally given by

$$V(x, y, z) = -ze^{2x} - yz^2 + \text{constant.}$$ (4.26)

To obtain the work done, we will utilise (4.17). It is given that the path of the particle is described by

$$\mathbf{r}(t) = (t, t, 1), \quad 0 \le t < 1.$$

Then

$$\dot{\mathbf{r}}(t) = (1, 1, 0)$$

and

$$\mathbf{F}(\mathbf{r}(t)) = (2e^{2t}, 1, 2t + e^{2t}).$$

Substituting these into (4.17) yields

$$W = \int_C \mathbf{F} \cdot d\mathbf{r} = \int_0^1 \mathbf{F}(\mathbf{r}(t)) \cdot \dot{\mathbf{r}}(t) dt$$

$$= \int_0^1 (2e^{2t} + 1) dt$$

$$= e^2 + 1 - 1 = e^2.$$

Alternatively, (useful if more complicated integrands arise) we can employ (4.19) directly and therefore dispense entirely with the integration process. Thus

$$W = \int_C F(\mathbf{r}(t)) \cdot d\mathbf{r} = - \int_C \nabla V(\mathbf{r}(t)) \cdot d\mathbf{r}$$
$$= V(\mathbf{r}(t_1)) - V(\mathbf{r}(t_2)). \tag{4.27}$$

Now, $\mathbf{r}(t_1) = \mathbf{r}(0) = (0,0,1)$ and $\mathbf{r}(t_2) = \mathbf{r}(1) = (1,1,1)$. Substituting these values into (4.27) yields

$$\begin{aligned} W &= V(0,0,1) - V(1,1,1) \\ &= (-1) - (-e^2 - 1) \quad \text{from (4.26)} \\ &= e^2, \end{aligned}$$

which is the same result as before. Notice that, without loss of generality, the constant in (4.26) has been put to zero; in fact, the choice of values for this constant is immaterial as we can choose any zero of the potential. □

4.3 Charged particles in an electromagnetic field

Particles possessing charge when subject to an *electromagnetic field* can behave in a rather interesting manner. We will first consider a charged particle in a magnetic field before proceeding to the full electromagnetic field.

4.3.1 Particle in a magnetic field

The equation of motion for a particle of mass m and charge q subject to an electromagnetic field is given by the *Lorentz force equation*:

$$\mathbf{F} = m\ddot{\mathbf{r}} = q(\mathbf{E}(\mathbf{r},t) + \dot{\mathbf{r}} \times \mathbf{B}(\mathbf{r},t)), \tag{4.28}$$

where $\mathbf{E}$ is the *electric field* and $\mathbf{B}$ is the *magnetic field*; both fields are position and time dependent. However, we will here consider only uniform electromagnetic fields that do not vary in space and time.

Assume then that our particle is subject to a uniform magnetic field that is parallel to the x-axis, $\mathbf{B} = (B,0,0)$, and $\mathbf{E} = \mathbf{0}$. From (4.28)

$$\begin{aligned} \ddot{\mathbf{r}} &= \frac{q}{m}(\dot{\mathbf{r}} \times \mathbf{B}) \\ &= \omega(\dot{\mathbf{r}} \times \mathbf{i}), \end{aligned}$$

where $\omega = qB/m$. On taking the cross product and comparing components, we find that

$$\ddot{x} = 0 \tag{4.29}$$

$$\ddot{y} = \omega\dot{z} \tag{4.30}$$

$$\ddot{z} = -\omega\dot{y}. \tag{4.31}$$

As (4.29) is not expressed in terms of y, z or their derivatives it can be treated independently of (4.30) and (4.31). Thus, integrating (4.29) twice yields

$$x(t) = v_x t + x_0, \tag{4.32}$$

where v_x and x_0 are constants of integration. On the other hand, (4.30) and (4.31) are *coupled* and cannot be directly integrated separately. However, we can use a trick to effectively uncouple these equations.

Differentiating (4.30) and substituting (4.31) into this yields a second-order differential equation in $\dot{y}$:

$$\dddot{y} = -\omega^2 \dot{y}. \tag{4.33}$$

This is merely a form of the simple harmonic motion equation given by (3.4) and the solution of this was given by (3.8). Hence, the solution of (4.33) is given by

$$\dot{y}(t) = A\cos(\omega t - \phi), \tag{4.34}$$

where A and ϕ are arbitrary constants. Substituting (4.34) into (4.30) yields

$$\dot{z}(t) = -A\sin(\omega t - \phi). \tag{4.35}$$

Finally, integrating (4.34) and (4.35) and combining these expressions with (4.32) gives the motion of the particle in the magnetic field:

$$x(t) = v_x t + x_0 \tag{4.36}$$

$$y(t) = \frac{A}{\omega}\sin(\omega t - \phi) + y_0 \tag{4.37}$$

$$z(t) = \frac{A}{\omega}\cos(\omega t - \phi) + z_0. \tag{4.38}$$

It follows from (4.37) and (4.38) that

$$(y - y_0)^2 + (z - z_0)^2 = \left(\frac{A}{\omega}\right)^2.$$

This implies that the particle is performing motion in a circle of radius A/ω and centre (y_0, z_0). (Note that we could dispense with y_0 and z_0 by shifting the origin to this point.) Equation (4.36) tells us that $\dot{x} = v_x =$ constant, so the particle is also travelling at a uniform velocity v_x in the x-direction. Hence, the particle is performing helical motion as depicted in Figure 4.4.

The speed of the particle along this path is $\sqrt{v_x^2 + A^2}$. If particle motion

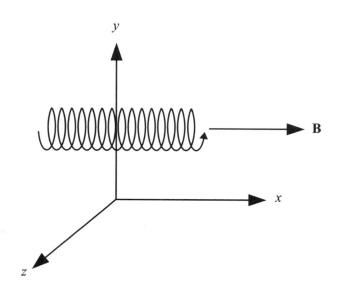

FIGURE 4.4
Motion of a particle in a magnetic field

is restricted to the yz-plane, $v_x = 0$, so that it traverses a circle of radius A/ω, then the particle will complete one revolution in a time $2\pi/\omega$.

The constant ω is sometimes referred to as the *cyclotron frequency*, a name associated with the *cyclotron accelerator* — a device used by physicists to accelerate charged elementary particles. As the particle travels through the magnetic field, a tangential electric field is applied to increase the speed of the particle. Each time the particle makes a circuit through the electric field, its speed increases by a small amount. The cyclotron frequency ω is independent of the speed of the particle. This means that as $|A|$ increases, the radius of the particle's path also increases until a maximum is reached and the particle is deflected through a thin window.

4.3.2   Crossed electric and magnetic fields

Consider a particle of mass m and charge q moving in a uniform electric field $\mathbf{E} = (E_x, 0, E_z)$ and magnetic field $\mathbf{B} = (B, 0, 0)$. From (4.28) the equations of motion are

$$\ddot{x} = \frac{q}{m} E_x \tag{4.39}$$

$$\ddot{y} = \frac{q}{m} B\dot{z} \tag{4.40}$$

$$\ddot{z} = \frac{q}{m}(E_z - B\dot{y}). \tag{4.41}$$

Equation (4.39) is constant on the right-hand side, and so it can be integrated in a straightforward manner to give

$$x(t) = x_0 + v_x t + \frac{1}{2}\frac{qE_x}{m}t^2, \tag{4.42}$$

where x_0 and v_x are constants of integration that can be determined from some given initial conditions. To solve the coupled equations (4.40) and (4.41), we will employ the technique used previously; namely, differentiate (4.40) and substitute in (4.41) yielding a second-order equation in $\dot{y}$:

$$\dddot{y} = -\left(\frac{qB}{m}\right)^2 \dot{y} + \frac{q^2 B}{m^2} E_z,$$

which can be written as

$$\frac{d^2 \dot{y}}{dt^2} + \omega^2 \dot{y} = \alpha\omega, \tag{4.43}$$

where $\omega = qB/m$ and $\alpha = qE_z/m$. Equation (4.43) is of the form of a simple harmonic motion equation with a forcing term $\alpha\omega$, so the solution is readily obtainable:

$$\dot{y}(t) = A\cos(\omega t - \phi) + \frac{\alpha}{\omega}, \tag{4.44}$$

where A and ϕ are arbitrary constants. Substituting (4.44) into (4.40) yields

$$\dot{z}(t) = -A\sin(\omega t - \phi). \tag{4.45}$$

Finally, integrating (4.44) and (4.45) and combining these expressions with (4.42) gives the motion of the particle in the electromagnetic field:

$$x(t) = \frac{1}{2}\alpha E_x t^2 + v_x t + x_0 \tag{4.46}$$

$$y(t) = \frac{A}{\omega}\sin(\omega t - \phi) + \frac{\alpha}{\omega}t + y_0 \tag{4.47}$$

$$z(t) = \frac{A}{\omega}\cos(\omega t - \phi) + z_0, \tag{4.48}$$

where y_0 and z_0 are arbitrary constants. For $E_z = 0$, it is clear from (4.47) and (4.48) that the particle will move in a circle of radius $\mid A/\omega \mid$, centre (y_0, z_0) and cyclotron frequency ω. E_y has the effect of combining uniform translatory motion in the x-direction with circular motion in the yz-plane. The ensuing motion in the yz-plane will be cycloidial.

4.4   Exercises

(In the following exercises take $g \approx 10 \text{ ms}^{-2}$.)

1. A projectile is launched at $t = 0$ with speed u from an origin O at
    an angle θ_1 to the horizontal. As the projectile moves freely under
    gravity it reaches its highest point along its path at A, where OA is
    inclined at an acute angle θ_2 to the horizontal. Find a trigonometric
    relationship between θ_1 and θ_2. Find a value for the speed u if
    $\theta = \pi/4$ and the vertical distance between O and A is 50 m. How
    much time has elapsed between the initial launch of the particle and
    the point where the velocity of the particle is parallel to OA?

2. Two projectiles, A and B, are launched simultaneously at $t = 0$ with
    speed u from an origin O at angles θ and $\pi - \theta$ to the horizontal,
    respectively ($\theta < \frac{\pi}{4}$). The projectiles move freely under gravity
    and in the same vertical plane. If Ox and Oy are, respectively, the
    horizontal and vertical (upward) axes in this plane, obtain the (x, y)
    positions of both particles at a later time t. Find the angle between
    the straight line joining these points and the x-axis, and obtain the
    point $(x_{\max}, y_{\max})$ of A — the maximum height along its path. At
    the instant when A reaches $(x_{\max}, y_{\max})$, show that the distance
    between A and B is

    $$\frac{\sqrt{2}}{20} u^2 (\sin 2\theta + \cos 2\theta - 1).$$

    For what value of θ is the distance between A and B greatest?

3. A projectile is launched up an inclined plane (making an angle θ_1
    to the horizontal) with a velocity u at an angle θ_2 to the plane. If
    the direction of projection lies in the vertical plane containing the
    line of greatest slope, find the time of flight and the range of the
    particle along its path. At the instant the projectile hits the plane
    it is travelling horizontally. Prove that

    $$\tan \theta_2 = \frac{\sin 2\theta_1}{3 - \cos 2\theta_1}.$$

4. A plane Π_1 is inclined at an angle α to the horizontal. A second
    horizontal plane Π_2 meets the first in a straight line. A projectile is
    launched with speed u from a point on Π_2 a distance d_2 from the
    line and at an angle β to the horizontal; the path of the projectile
    is in a plane perpendicular to the line. If the projectile hits Π_1 at a
    distance d_1 from the line, show that

    $$d_1 u^2 \sin \alpha = (d_2 + d_1 \cos \alpha)(u^2 \tan \beta - 5(d_2 + d_1 \cos \alpha) \sec^2 \beta).$$

Provided that the root exists, prove that for some u the maximum value of d_1 is the positive root of the equation

$$u^4 - 20d_1 u^2 \sin\alpha - 100(d_2 + d_1 \cos\alpha) = 0.$$

Show, also, that if $u^2 < 10d_2$ the projectile will be unable to reach Π_2.

5. A projectile of mass m is launched with velocity u at an angle θ to the horizontal. Given that the projectile is subject to gravity and a horizontal air resistance mkv, where v is the horizontal component of the projectile's velocity, show that the range R on the horizontal plane through the point of projection is

$$R = \frac{u}{k}\left(1 - \exp\left(-\frac{ku}{5}\sin\theta\right)\right)\cos\theta.$$

Show, also, that the horizontal distance x travelled by the projectile as it reaches its maximum height is $x > R/2$.

6. The equation of motion of a projectile under the influence of a gravitational field is given by

$$\ddot{\mathbf{r}} = \mathbf{g},$$

where $\mathbf{g}$ is the uniform constant gravitational field. If the projectile passes the points with position vectors $\mathbf{r_1}, \mathbf{r_2}$ and $\mathbf{r_3}$ at times t_1, t_2 and t_3, respectively, derive the result

$$\mathbf{g} = -\frac{2(t_2 - t_3)\mathbf{r_1} + (t_3 - t_1)\mathbf{r_2} + (t_1 - t_2)\mathbf{r_3}}{(t_2 - t_3)(t_3 - t_1)(t_1 - t_2)}.$$

7. A cannonball of mass m is fired into the wind at $t = 0$, from an origin at ground level, with velocity $\mathbf{v}$ and at an angle of elevation $0 < \theta < \pi/2$. The cannonball moves under a gravitational field $\mathbf{g} = (0, -g, 0)$ but its motion is opposed by a resistance proportional to $\mathbf{v} - \mathbf{w}$, where $\mathbf{w}$ is the velocity of the wind that opposes the motion of the cannonball. Show that the position of the cannonball at a subsequent time t is given by

$$\mathbf{r}(t) = \left(t - \frac{m(1 - e^{-(k/m)t})}{k}\right)\mathbf{w} + m\frac{(1 - e^{-(k/m)t})}{k}\mathbf{v}$$
$$+ \frac{m}{k}\left(t - \frac{(1 - e^{-(k/m)t})}{k}\right)\mathbf{g},$$

where k is a constant of proportionality. The cannonball is now fired at an angle of elevation $\pi/2$ into a wind blowing perpendicularly to its path. Show that the time of flight τ satisfies

$$\tau = \frac{m}{k}\ln\left[\frac{kv + gm}{kv + g(m - k\tau)}\right],$$

where $v = |\mathbf{v}|$.

8. Determine the work done in moving a particle along a straight line that joins the points $(0,0,0)$ and $(2,1,3)$ in a force field $\mathbf{F} = (3x^2, 2xz - y, z)$.

9. Show that $\mathbf{F} = (y^2 \cos x + z^3, 2y \sin x - 4, 3xz^2 + 2)$ is a conservative field and find the scalar potential energy for $\mathbf{F}$. What is the work done in moving a particle along a straight line joining the points $(0, 1, -1)$ to $(\pi/2, -1, 2)$?

10. A particle of mass m and charge q moves in a uniform electromagnetic field such that $\mathbf{E} = (0, E, 0)$ and $\mathbf{B} = (0, 0, B)$. If the particle is initially moving in the xy-plane at $t = 0$, show that it will remain in the xy-plane at all subsequent times. Furthermore, show that the particle moves in a circle with a constant drift in the direction $\mathbf{E} \times \mathbf{B}$ and find the position of the particle at a later time t in terms of E, B and ω, where $\omega = qB/m$.

11. A particle of mass m and charge q moves in a uniform magnetic field $\mathbf{B} = (B, 0, 0)$ and a gravitational field $\mathbf{g} = (0, -g, 0)$. Show that the motion of the particle is helical combined with a uniform horizontal drift.

5

Central Forces and Orbits

CONTENTS

Many important problems in physics are concerned with the motion of a particle under the action of a *central force*. Two specific examples of this generic force that we will be considering are the gravitational (attractive) force and the electrostatic (repulsive) force. Both forces are governed by an *inverse square law*; in the gravitational case this was defined by (1.13). The term *orbit* is used to describe the path of a particle subject to these central forces.

5.1 Central forces and angular momentum

A force $\mathbf{F}$ acting on a particle P is defined as a central force if the magnitude F of the force is a function only of the distance r of P from a fixed origin O; and furthermore, $\mathbf{F}$ is directed along a straight line connecting O to P, the direction of this line being $\hat{\mathbf{r}}$, a unit vector in the direction of $\mathbf{r}$ (see Figure 5.1).

Thus

$$\mathbf{F}(\mathbf{r}) = F(r)\hat{\mathbf{r}}, \tag{5.1}$$

where $F = |\mathbf{F}|$, $r = |\mathbf{r}|$ and $\hat{\mathbf{r}} = \mathbf{r}/r$. In the case when $F(r) > 0$ the force

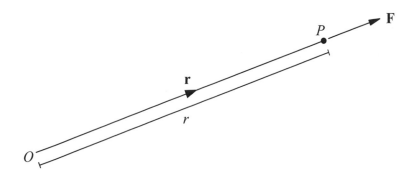

FIGURE 5.1
Particle in a central force

points radially out from the origin and is said to be *repulsive*. Conversely, if $F(r) < 0$ the force points in the direction of the origin, and it is said to be *attractive*.

Consider now the particle P with mass m under the influence of a central force $\mathbf{F}$. Assuming no other forces are present and the frame containing the fixed origin O is inertial, using Newton's second law, we can write the equation of motion of P as

$$m\ddot{\mathbf{r}} = F(r)\hat{\mathbf{r}}. \tag{5.2}$$

Next, we define the *angular momentum* (or *moment of momentum*) $\mathbf{L}$ of P about the origin O at position $\mathbf{r}$:

$$\mathbf{L} = m\mathbf{r} \times \dot{\mathbf{r}} = \mathbf{r} \times \mathbf{p}, \tag{5.3}$$

where $\mathbf{p} = m\dot{\mathbf{r}}$ is the linear momentum of P. Notice that the angular momentum acts in a direction perpendicular to both the position vector $\mathbf{r}$ and linear momentum vector $\mathbf{p}$ as depicted in Figure 5.2.

That the angular momentum of P is conserved under the action of a central force can be seen by differentiating (5.3) with respect to time. Thus

$$\begin{aligned}
\dot{\mathbf{L}} \equiv \frac{d\mathbf{L}}{dt} &= \frac{d}{dt}(m\mathbf{r} \times \dot{\mathbf{r}}) \\
&= m(\dot{\mathbf{r}} \times \dot{\mathbf{r}}) + m(\mathbf{r} \times \ddot{\mathbf{r}}) \\
&= \mathbf{0} + m(\mathbf{r} \times \ddot{\mathbf{r}}) \\
&= \mathbf{r} \times \mathbf{F}.
\end{aligned}$$

The quantity $\mathbf{r} \times \mathbf{F}$ is called the *torque* (or *moment* of the force) and is designated here by the symbol $\mathbf{G}$. Thence,

$$\dot{\mathbf{L}} = \mathbf{G} = \mathbf{r} \times \mathbf{F}. \tag{5.4}$$

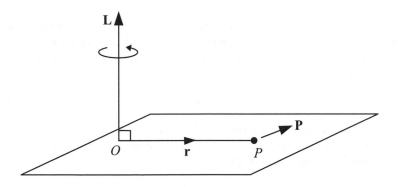

FIGURE 5.2
Angular momentum of P. Note that $\mathbf{r}$ and $\mathbf{p}$ are coplanar

It follows that the torque acts in a direction perpendicular to both $\mathbf{r}$ and $\mathbf{F}$ (see Figure 5.3), and can be considered as defining the axis of rotation

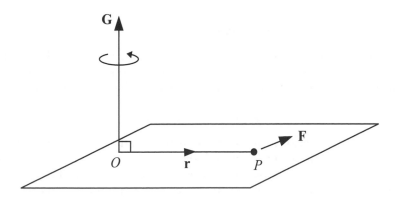

FIGURE 5.3
Torque perpendicular to the plane containing $\mathbf{r}$ and $\mathbf{F}$

about which P rotates subject to $\mathbf{F}$. One should briefly consider the rather important equation (5.4). An obvious analogy can be made between it and Newton's second law $\dot{\mathbf{p}} = \mathbf{F}$ for the rate of change of linear momentum. As $\mathbf{F}$ is being considered as a central force and, therefore, parallel to $\mathbf{r}$, the torque must vanish, leaving

$$\dot{\mathbf{L}} = \mathbf{0}. \tag{5.5}$$

Hence, $\mathbf{L} = $ constant and angular momentum is conserved throughout the motion.

An important consequence of our discussion is that the motion of P is restricted to a two-dimensional subspace of our Euclidean 3-space $\mathbb{E}^3$; namely, the plane containing $\mathbf{r}$ and $\dot{\mathbf{r}}$. This can be easily verified by taking the scalar product of (5.3) with $\mathbf{r}$:

$$\mathbf{L} \cdot \mathbf{r} = m(\mathbf{r} \times \dot{\mathbf{r}}) \cdot \mathbf{r} = 0. \tag{5.6}$$

The angular momentum being fixed in $\mathbb{E}^3$ is always perpendicular to the position vector $\mathbf{r}$; consequently, the motion of P takes place in a plane that is always perpendicular to $\mathbf{L}$ (see Figure 5.2).

5.2   Circular motion

Now that we have established that the particle P under the influence of a central force with centre O moves in a plane that passes through O, from (5.6), it will prove convenient (especially when discussing orbits) to model the dynamical behaviour of P with the aid of plane polar coordinates (r, θ).

5.2.1   Plane polar coordinates

Let us begin by orientating our Cartesian coordinate system such that the motion of P is constrained in the xy-plane and, therefore, the angular momentum vector is pointing in the z-axis. On transforming to polar coordinates, we have

$$x = r\cos\theta, \quad y = r\sin\theta,$$

where θ is the angle between the position vector $\mathbf{r}$ and the x-axis as depicted in Figure 5.4. The unit vectors $\hat{\mathbf{r}}$ and $\hat{\boldsymbol{\theta}}$,[1] shown in Figure 5.4, point in the direction of increasing r and θ, respectively, and are non-constant. Like the unit vectors $\mathbf{i}$ and $\mathbf{j}$ in the Cartesian coordinate system, $\hat{\mathbf{r}}$ and $\hat{\boldsymbol{\theta}}$ form an orthonormal basis at each point in the plane. It is, therefore, a straightforward task to transform between both coordinate systems. Hence, using the terminology

$$\mathbf{r} = r\cos\theta\,\mathbf{i} + r\sin\theta\,\mathbf{j} \equiv (r\cos\theta, r\sin\theta),$$

we have

$$\hat{\mathbf{r}} = (\cos\theta, \sin\theta) \quad \text{and} \quad \hat{\boldsymbol{\theta}} = (-\sin\theta, \cos\theta).$$

Notice in Figure 5.4 that as P moves in the plane, $\hat{\mathbf{r}}$ and $\hat{\boldsymbol{\theta}}$ will depend on θ but not on r. Consequently, $\hat{\mathbf{r}}$ and $\hat{\boldsymbol{\theta}}$ are said to be *vector-valued* functions of

---

[1]The unit vectors $\hat{\mathbf{r}}$ and $\hat{\boldsymbol{\theta}}$ are sometimes referred to as the *radial unit vector* and *transverse unit vector*, respectively.

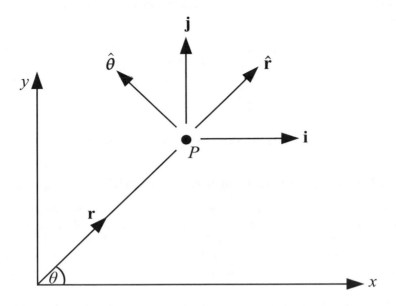

FIGURE 5.4
Plane polar coordinates

θ and are written as $\hat{\mathbf{r}} = \hat{\mathbf{r}}(\theta)$ and $\hat{\boldsymbol{\theta}}=\hat{\boldsymbol{\theta}}(\theta)$. Thus, as P moves in the transverse direction, we have

$$\frac{d\hat{\mathbf{r}}}{d\theta} = (-\sin\theta, \cos\theta) = \hat{\boldsymbol{\theta}},$$

$$\frac{d\hat{\boldsymbol{\theta}}}{d\theta} = (-\cos\theta, -\sin\theta) = -\hat{\mathbf{r}}.$$

Assuming P moves in such a way that θ and r are functions of time, we have, on employing the chain rule,

$$\frac{d\hat{\mathbf{r}}}{dt} = \frac{d\theta}{dt}\frac{d\hat{\mathbf{r}}}{d\theta} = \dot{\theta}\hat{\boldsymbol{\theta}}$$

$$\frac{d\hat{\boldsymbol{\theta}}}{dt} = \frac{d\theta}{dt}\frac{d\hat{\boldsymbol{\theta}}}{d\theta} = -\dot{\theta}\hat{\mathbf{r}}.$$

Writing the position vector of P as

$$\mathbf{r} = r\hat{\mathbf{r}}$$

and differentiating with respect to time yields

$$\dot{\mathbf{r}} = \dot{r}\hat{\mathbf{r}} + r\frac{d\hat{\mathbf{r}}}{dt}$$

$$= \dot{r}\hat{\mathbf{r}} + r\dot{\theta}\hat{\boldsymbol{\theta}}, \tag{5.7}$$

which is the velocity of P in the plane. The acceleration of P can be found by differentiating (5.7) with respect to time:

$$\ddot{\mathbf{r}} = \ddot{r}\hat{\mathbf{r}} + \dot{r}\frac{d\hat{\mathbf{r}}}{dt} + \dot{r}\dot{\theta}\hat{\boldsymbol{\theta}} + r\ddot{\theta}\hat{\boldsymbol{\theta}} + r\dot{\theta}\frac{d\hat{\boldsymbol{\theta}}}{dt}$$
$$= \ddot{r}\hat{\mathbf{r}} + \dot{r}\dot{\theta}\hat{\boldsymbol{\theta}} + \dot{r}\dot{\theta}\hat{\boldsymbol{\theta}} + r\ddot{\theta}\hat{\boldsymbol{\theta}} - r\dot{\theta}^2\hat{\mathbf{r}}$$
$$= (\ddot{r} - r\dot{\theta}^2)\hat{\mathbf{r}} + (r\ddot{\theta} + 2\dot{r}\dot{\theta})\hat{\boldsymbol{\theta}}. \tag{5.8}$$

The following sections illustrate some of the ideas presented here.

5.2.2   Motion of a particle in a circular orbit

The simplest type of orbital path along which a particle can traverse is a circle. However, a particle subject to the gravitational field of another body will never, in practice, achieve a perfectly circular orbit about that body — a certain amount of orbital eccentricity will be present. Nevertheless, to a very good approximation,[2] planetary motion about the Sun may be considered as circular.

Consider a particle P of mass m under the influence of a gravitational force $\mathbf{F}$ of a fixed body with mass M situated at an origin O. As we have already stipulated that the orbit is circular, the distance $OP =$ constant. Writing this constant as R, we have $r = R$ and, therefore, $\dot{r} = 0$. Thence, from (5.7),

$$\dot{\mathbf{r}} = R\dot{\theta}\hat{\boldsymbol{\theta}}, \tag{5.9}$$

which is the velocity of P in the plane of motion. Notice that it acts only in the transverse direction. Employing (5.8), the acceleration of P is given by

$$\ddot{\mathbf{r}} = -R\dot{\theta}^2\hat{\mathbf{r}} + R\ddot{\theta}\hat{\boldsymbol{\theta}}. \tag{5.10}$$

However, the gravitational force is central and therefore acts along a line parallel to $\mathbf{r}$ (see (5.2)). The equation of motion of P is then, using (5.2) and (1.13),

$$\mathbf{F} = m\ddot{\mathbf{r}} = -\frac{GmM}{R^2}\hat{\mathbf{r}}. \tag{5.11}$$

On comparing components of $\hat{\mathbf{r}}$ and $\hat{\boldsymbol{\theta}}$ in (5.10) and (5.11), we find that

$$R\dot{\theta}^2 = \frac{GmM}{R^2} \tag{5.12}$$

and

$$R\ddot{\theta} = 0. \tag{5.13}$$

---

[2]For example, the current value of Earth's eccentricity is about 0.00167, but is subject to change due to the gravitational attractions of other planets.

Equation (5.13) implies that the quantity $\dot{\theta}$, called the *angular speed*, is constant; that is, $\dot{\theta} \equiv \omega =$ constant. Thus, the speed of P in the plane is, by (5.9),

$$|\dot{\mathbf{r}}| = v = R\omega$$

or, by (5.12),

$$v = \sqrt{\frac{GM}{R}}.$$

The magnitude of the acceleration is then

$$|\ddot{\mathbf{r}}| = a = R\omega^2$$

or

$$a = \frac{GM}{R^2}.$$

5.2.3   Motion of a particle in a vertical circle

Consider a particle P of mass m attached to a light inextensible string of length l and the other end fixed at an origin O (Figure 5.5). This particular arrangement is that of a *simple pendulum*.

Initially, the particle is allowed to hang stationary at A before being given a horizontal speed u. Some time later the particle arrives at B with speed v. It is assumed that the string is taut, f_T being the tension in the string, so that the path forms an arc AB of a vertical circle. The tension in the string does no work due to the string being inextensible; therefore, the change in kinetic energy $\frac{m}{2}(v^2 - u^2)$ is due only to the work done in moving the particle from A to B, namely, $-mgl(1 - \cos\theta)$. Hence, the conservation of energy tells us that

$$\frac{m}{2}(v^2 - u^2) = -mgl(1 - \cos\theta)$$

or

$$v^2 = u^2 - 2gl(1 - \cos\theta). \tag{5.14}$$

The tension in the string can be calculated with the aid of (5.8). Thence,

$$\ddot{\mathbf{r}} = -l\dot{\theta}^2\hat{\mathbf{r}} + l\ddot{\theta}\hat{\boldsymbol{\theta}}$$

giving the equation of motion of the particle as

$$ml(-\dot{\theta}^2\hat{\mathbf{r}} + \ddot{\theta}\hat{\boldsymbol{\theta}}) = -mg\mathbf{j} - f_T\hat{\mathbf{r}}$$

or, after resolving $\mathbf{j}$,

$$ml(-\dot{\theta}^2\hat{\mathbf{r}} + \ddot{\theta}\hat{\boldsymbol{\theta}}) = -mg(\sin\theta\hat{\boldsymbol{\theta}} + \cos\theta\hat{\mathbf{r}}) - f_T\hat{\mathbf{r}}.$$

On equating radial and transverse components, we have

$$-ml\dot{\theta}^2 = mg\cos\theta - f_T \tag{5.15}$$

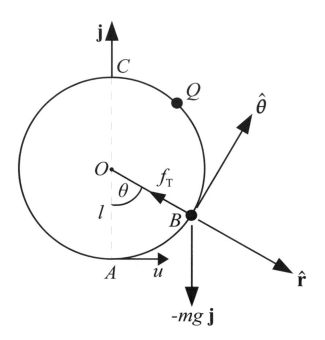

FIGURE 5.5
Motion of a particle in a vertical circle

and

$$ml\ddot{\theta} = -mg\sin\theta. \tag{5.16}$$

It will be more useful for our discussion to rewrite (5.15) with the aid of (5.14) and $v = l\dot{\theta}$, as

$$f_{\mathrm{T}} = \frac{m}{l}(u^2 - gl(2 - 3\cos\theta)). \tag{5.17}$$

We can now analyse the dynamical behaviour of the particle in four different cases for a given initial speed u.

Case (1): $u^2 < 2gl$
The speed of the particle vanishes when $\cos\theta = 1 - u^2/2gl$ from (5.14), but $f_{\mathrm{T}} = m(gl - u^2/2)/l$ from (5.17) and so the tension is non-zero. The particle will perform oscillatory motion about A by first of all travelling along the arc AB and then along an equivalent distance on the other side of A. Throughout the oscillatory motion, the speed of the particle as it passes A will be u, from (5.14).

Case (2): $u^2 = 2gl$
Both the speed of the particle and the tension in the string vanish when $\theta = \pi/2$. The particle performs oscillatory motion about A, the path being that of a semicircle.

Case (3): $2gl < u^2 < 5gl$
As the particle travels along its path, f_T decreases as θ increases according to (5.17). At a point Q for which $\pi/2 < \theta < \pi$, f_T vanishes completely when $\cos\theta = \frac{2}{3}(1 - u^2/2g)$; the corresponding speed is given by $v^2 = \frac{1}{3}(u^2 - 2gl)$. Thus, at Q the tension in the string vanishes and the particle falls under gravity along a parabolic path until the string becomes taut.

Case (4): $5gl \leq u^2$
As θ increases the speed of the particle and the tension in the string will be positive along the particle's path. At the point C $(\theta = \pi)$ we have $v^2 = u^2 - 4gl$ and $f_T = m(u^2 - 5gl)/l$. As both v and f_T will remain positive when the particle passes C, it follows that complete circles will be performed.

If the particle performs small oscillations, $|\theta| \ll 1$, then we can use the approximation $\sin\theta \approx \theta$ (θ in radians) to write (5.16) as

$$\ddot{\theta} + \omega^2\theta = 0,$$

where $\omega = \sqrt{g/l}$. One will recognise this as the equation of simple harmonic motion (see (3.4)). The general solution is then

$$\theta(t) = A\cos(\omega t - \psi),$$

where given initial conditions can be used to fix the arbitrary constants A and ψ. The period of small oscillations is given by

$$\tau = \frac{2\pi}{\omega} = 2\pi\sqrt{\frac{l}{g}},$$

which is the approximate period of a simple pendulum. Notice that this period is independent of the amplitude of swing, A. This is an important property that is exploited in many pendulum devices.

5.3 Orbital motion

For a particle P with mass m subject to a central force $\mathbf{F}$, it was established in Section 5.1 that the equation of motion of P is given by (5.2). The acceleration of P, in plane polar coordinates, was given by (5.8). On substituting (5.8) into (5.2), the equation of motion of P becomes

$$m(\ddot{r} - r\dot{\theta}^2)\hat{\mathbf{r}} + m(r\ddot{\theta} + 2\dot{r}\dot{\theta})\hat{\boldsymbol{\theta}} = F(r)\hat{\mathbf{r}}$$

or, more conveniently,

$$m(\ddot{r} - r\dot{\theta}^2)\hat{\mathbf{r}} + \frac{m}{r}\frac{d}{dt}(r^2\dot{\theta})\hat{\boldsymbol{\theta}} = F(r)\hat{\mathbf{r}}.$$

On comparing radial and transverse components, we have

$$m(\ddot{r} - r\dot{\theta}^2) = F(r) \tag{5.18}$$

$$\frac{d}{dt}(mr^2\dot{\theta}) = 0. \tag{5.19}$$

Integrating (5.19) yields $mr^2\dot{\theta} = $ constant. Clearly, the quantity $mr^2\dot{\theta}$ is conserved, and although it is not obvious, this quantity is in fact the magnitude of angular momentum. To prove this, merely substitute (5.7) into (5.3) and take the modulus of the resulting equation:

$$\mathbf{L} = m\mathbf{r} \times \dot{\mathbf{r}} = mr\hat{\mathbf{r}} \times (\dot{r}\hat{\mathbf{r}} + r\dot{\theta}\hat{\boldsymbol{\theta}})$$
$$= mr^2\dot{\theta}(\hat{\mathbf{r}} \times \hat{\boldsymbol{\theta}}),$$

therefore,

$$L \equiv |\mathbf{L}| = mr^2\dot{\theta}. \tag{5.20}$$

As $\hat{\mathbf{r}} \times \hat{\boldsymbol{\theta}}$ is a unit vector, its modulus is one. Because the central force $\mathbf{F}$ given by (5.1) is conservative, we can write down the following form for the total energy E using conservation of energy $E = T + V$:

$$E = \frac{1}{2}m\dot{r}^2 + \frac{1}{2}mr^2\dot{\theta}^2 + V(r), \tag{5.21}$$

where the first two terms on the right represent the total kinetic energy T and the potential energy is $V(r)$, where $-dV(r)/dr = F(r)$. Substituting the expression for angular momentum (5.20) into (5.21) yields

$$E = \frac{1}{2}m\dot{r}^2 + \frac{L^2}{2mr^2} + V(r). \tag{5.22}$$

Due to the fact that (5.22) has no θ-dependence, it is sometimes referred to as the *radial energy equation*. Rearranging for $\dot{r}$, the resulting integral can be evaluated for $r(t)$. Similarly, (5.20) can be integrated and evaluated for $\theta(t)$. Thence, the solutions of these equations can be given in terms of E and L together with some specified initial conditions.

   If we compare (5.22) with the one-dimensional conservation of energy equation (2.3) or (2.4) an analogy can be clearly drawn between them: the kinetic energy term in (5.22) replaces $\dot{x}$ with $\dot{r}$, and the potential energy term replaces $V(r)$ with $V_{\text{eff}}(r)$, where

$$V_{\text{eff}}(r) = \frac{L^2}{2mr^2} + V(r). \tag{5.23}$$

The term *effective potential* is often employed to represent $V_{eff}(r)$ as it effectively reduces the type of motion of P from radial motion to rectilinear motion. To further elucidate, consider the equation obtained by substituting (5.20) into (5.18):

$$m\ddot{r} = F(r) + \frac{L^2}{mr^3}. \tag{5.24}$$

This is the *radial equation of motion* for P and it is analogous to a one-dimensional equation of motion for a particle under the influence of a force $F(r)$ combined with an additional force L^2/mr^3, referred to as the *centrifugal force*.[3] Integrating (5.24) yields

$$V_{eff}(r) = V(r) + \frac{L^2}{2mr^2},$$

which is (5.23). In integrating (5.24), we have used the fact that

$$m\ddot{r} = -\frac{dV_{eff}}{dr}$$

and

$$F(r) = -\frac{dV}{dr}.$$

The potential energy term $L^2/2mr^2$ is referred to as the *centrifugal barrier*. This is because the centrifugal force is repulsive and, therefore, P is prevented from getting too near to $r = 0$.

One should make no mistake about the dimensionality of the motion under discussion. The motion of P takes place in the plane and is therefore two-dimensional; however, we have been considering radial motion (dependent on the radial coordinate r) and in this context the motion is one-dimensional.

5.4 The inverse square law

Rearranging (5.22) for $\dot{r}$ and integrating yields

$$t - t_0 = \pm \left(\frac{m}{2}\right)^{\frac{1}{2}} \int_{r_0}^{r} \frac{dr'}{(E - V(r') - L^2/2m(r')^2)^{\frac{1}{2}}}$$

$$= \pm \left(\frac{m}{2}\right)^{\frac{1}{2}} \int_{r_0}^{r} \frac{dr'}{(E - V_{eff}(r'))^{\frac{1}{2}}}. \tag{5.25}$$

---

[3] The centrifugal force is considered among a group of forces called *fictitious forces*. They are introduced to preserve the integrity of Newton's second law of motion when dealing with non-inertial frames of reference. However, this force is manifest to anyone travelling in a car or bus as it rounds a bend — the outward push that one experiences is due to the centrifugal force.

This is analogous to (2.8) for the case of one-dimensional motion. Rather than attempting to solve (2.8) explicitly, we used a graphical approach to determine salient features of a particle's motion. We will again employ this approach, but here the effective potential energy V_{eff} will be plotted against r.

Consider the potential energy

$$V(r) = -\frac{k}{r} \tag{5.26}$$

for a central inverse square law of force

$$\mathbf{F} = -\frac{k}{r^2}\hat{\mathbf{r}}, \tag{5.27}$$

where k is a constant, that may be positive, negative or zero. The potential energy diagram for V_{eff} in (5.29) is shown in Figure 5.6. Let us consider in turn the three cases in Figure 5.6.

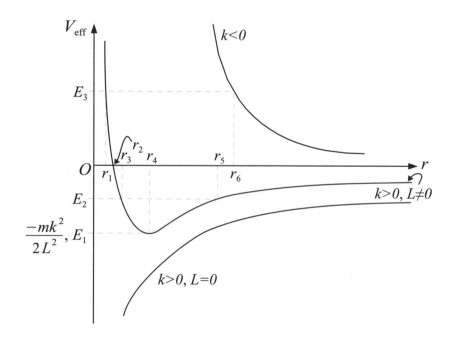

FIGURE 5.6
The potential diagram for V_{eff} given by (5.29)

(It may prove useful to refer to Section 2.2 as we discuss each of the three cases.)

The radial energy equation is then given as

$$E = \frac{1}{2}m\dot{r}^2 + V_{\text{eff}}(r), \tag{5.28}$$

where the effective potential energy V_{eff} is given by

$$V_{\text{eff}} = -\frac{k}{r} + \frac{L^2}{2mr^2}.$$ (5.29)

Case (1): $k < 0$
The effective potential $V_{\text{eff}}(r)$ is a repulsive potential and the only possible total energy is $E > 0$. Because $V_{\text{eff}}(r)$ is decreasing monotonically with increasing radial distance r, there is no possibility of periodic motion and circular motion cannot occur. A particle travelling from infinity will eventually reach a minimum radial distance r_6 for the given energy E_3. At this distance the radial speed is zero and the particle's speed is in the transverse direction. The particle then moves out towards infinity again. The orbit is *unbounded* and *hyperbolic*.

Case (2): $k > 0$, $L \neq 0$
The motion of a particle is characterised by four possible values of the total energy E as follows:

(i) $E = E_1 = -mk^2/2L^2$. The particle is located at the lowest point of the effective potential at $r = r_4$. It is, therefore, fixed in the radial direction indefinitely and so $\dot{r} = 0$. However, the motion is two-dimensional and the particle will have an angular speed $\dot{\theta} = L/r_4^2$. Thus the particle is performing motion in a *circle* of radius r_4, and the orbit is *bounded*.

(ii) $E = E_2$, $E_1 < E_2 < 0$. The particle is located between two radial distances: a minimum distance r_3 and a maximum distance r_5. As the particle moves towards r_5, $\dot{r} > 0$, and on reaching r_5, $\dot{r} = 0$. The particle then begins to move towards r_3 with $\dot{r} < 0$, and on reaching r_3, $\dot{r} = 0$. The cycle is then repeated indefinitely. The orbit is *bounded* and will be seen to be *elliptic*.
   If the particle had been a satellite orbiting the Earth, say, then the minimum distance r_3 represents the closest approach to the Earth and is called *perigee* (or *periapsis*), and the maximum distance r_5 represents the farthest retreat from the Earth, and is called *apogee* (or *apoapsis*). The corresponding terms for orbits around the Sun are *perihelion* and *aphelion*, respectively. The positions r_3 and r_5 are more generally known as *apsides*; and any point for which the radial speed is zero is known as an *apse*.

(iii) $E = 0$. The particle's closest approach is $r_2 = L^2/2mk$; however, the particle's energy is just sufficient for it to travel out to infinity. So, as $r \to \infty$, $\dot{r} \to 0$ and the kinetic energy approaches zero. The orbit is *unbounded* and will be seen to be *parabolic*.

(iv) $E = E_3 > 0$. The particle's closest approach is r_1 and again, as in the case of $E = 0$, the particle has enough energy to travel to infinity. The orbit is *unbounded* and and will be seen to be *hyperbolic*.

Case (3): $k > 0$, $L = 0$
The particle's motion is that of the one-dimensional motion discussed in Section 2.6.

5.5   The orbital equation

The procedure discussed above allows us to determine all the salient information concerning an orbiting particle's motion in the radial and transverse directions as a function of time. However, evaluating the integral (5.25) is extremely laborious for all but a very few simple instances of V_{eff}. Moreover, knowledge of $r = r(\theta)$ will, for many situations, be preferable than that for $r = r(t)$. Thus our aim is to derive an equation the solutions for which will give us the orbital paths of a particle subject to a given central force.

To achieve this, it is first useful to introduce a new dependent variable

$$u = \frac{1}{r}. \tag{5.30}$$

The purpose of this is to allow us to combine (5.18) and (5.20) into a single (and as we will see) second-order differential equation in u and θ. To begin with, we use the chain rule to rewrite $\dot{r}$:

$$\dot{r} \equiv \frac{dr}{dt} = \frac{dr}{d\theta}\frac{d\theta}{dt} = \frac{d}{d\theta}\left(\frac{1}{u}\right)\dot{\theta} = -\frac{1}{u^2}\frac{du}{d\theta}\dot{\theta}$$

$$= -r^2\dot{\theta}\frac{du}{d\theta} \quad \text{from (5.30)}$$

$$= -\frac{L}{m}\frac{du}{d\theta} \quad \text{from (5.20).} \tag{5.31}$$

Next, on using (5.20), (5.30) and (5.31),

$$\ddot{r} \equiv \frac{d}{dt}\left(\frac{dr}{dt}\right) = \frac{d}{d\theta}\left(\frac{dr}{dt}\right)\frac{d\theta}{dt} = \frac{d\dot{r}}{d\theta}\dot{\theta}$$

$$= -\frac{L^2 u^2}{m^2}\frac{d^2 u}{d\theta^2}. \tag{5.32}$$

Substituting (5.30) and (5.32) into (5.24) yields (after a little algebra)

$$\frac{d^2 u}{d\theta^2} + u = -\frac{mF(1/u)}{L^2 u^2}, \tag{5.33}$$

which is the required differential equation for an orbiting particle. Equation

(5.33) is called the *orbital* or *path equation*; its solutions will be plane curves, and in the case of the inverse square law of force these curves will be different types of conic sections. Notice that due to the u dependency on the right of (5.33) the orbital equation is in general non-linear. However, it so happens that for the inverse square law of force, the equation reduces to linear form.[4]

5.5.1 Orbital paths

The inverse square law of force is by far the most important force in orbit theory, and as the force also governs the scattering of oppositely charged particles in Rutherford scattering, it will be considered exclusively.

For the attractive gravitational field, the potential is given by (5.26), and hence, the force is given by (5.27). The orbital equation is thus

$$\frac{d^2u}{d\theta^2} + u = \frac{mk}{L^2}, \qquad (5.34)$$

where $k = GMm$. Equation (5.34) should be familiar as the equation of simple harmonic motion with a non-zero constant on the right. This second-order inhomogeneous equation is easily soluble and yields the following general solution[5]:

$$u(\theta) = A\cos(\theta - \theta_0) + \frac{mk}{L^2}, \qquad (5.35)$$

where A and θ_0 are constants of integration. Because θ_0 is arbitrary, we can, without loss of generality, choose $A \geq 0$. Furthermore, u is largest when r is smallest (see 5.30) — for a particle orbiting the Earth this corresponds to the perigee — so from (5.35), we have $u = A + mk/L^2$. If the major axis is set to coincide with the axis $\theta = 0$, the constant θ_0 — which merely orientates the orbit with respect to the coordinate axes — will vanish. Thus (5.35) can now be written with the aid of (5.30) as

$$\frac{1}{r} = \frac{1}{l}(1 + e\cos\theta), \qquad (5.36)$$

where $e = AL^2/mk$ and $l = L^2/mk$. Equation (5.36) is the polar equation of a *conic section* (see Appendix for an introduction to conic sections). The non-negative constant e is called the *eccentricity* and determines the particular type of conic section, and hence the orbit, under consideration, while the constant l is called the *semi-latus rectum* ($2l$ being the *latus rectum*).

There are three cases of eccentricity that require characterising: the ellipse ($0 \leq e < 1$), the hyperbola ($e > 1$), and the parabola ($e = 1$). We will consider each in turn.

---

[4]The only other case for which this is a linear equation is the inverse cube law.

[5]The general solution $u(\theta) = A\cos\theta + B\sin\theta + mk/L^2$ may appear more natural, but the form given by (5.35) is better suited for defining orbits as conic sections.

Elliptic orbits $0 \leq e < 1$

The fact that this condition characterises an ellipse from (5.36) is most readily seen by reverting to Cartesian coordinates with $x = r\cos\theta$ and $y = r\sin\theta$. Thus, on rearranging (5.36),

$$r = l - er\cos\theta,$$

and writing in terms of Cartesian coordinates

$$x^2 + y^2 = (l - ex)^2,$$

we find, after a little algebra,

$$\frac{(x + ae)^2}{a^2} + \frac{y^2}{b^2} = 1, \tag{5.37}$$

where

$$a^2 = \frac{l^2}{(1 - e^2)^2} \qquad \text{and} \qquad b^2 = \frac{l^2}{1 - e^2}. \tag{5.38}$$

Equation (5.37) describes an ellipse with centre C at $(-ae, 0)$. From (5.38), we can see that $a^2 > b^2$; hence the semi-major axis is a and the semi-minor axis is b (see Figure 5.7).

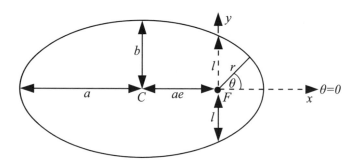

FIGURE 5.7
The elliptic orbit

In terms of the semi-major axis, the polar form of (5.37) is, by (5.36),

$$r = \frac{a(1 - e^2)}{1 + e\cos\theta}. \tag{5.39}$$

For a particle orbiting a planet positioned at a focus, F, of the ellipse, a simple calculation gives the perigee and apogee as $a(1 - e)$ and $a(1 + e)$, respectively. For $e = 0$ the focus and centre coincide resulting in a circular orbit. The semi-latus rectum l would then correspond to the orbital radius.

Hyperbolic orbits $e > 1$

The calculation leading to (5.37) can essentially be repeated to yield the following equation describing a hyperbola with focus F at $(-ae, 0)$:

$$\frac{(x - ae)^2}{a^2} - \frac{y^2}{b^2} = 1, \qquad (5.40)$$

where

$$a^2 = \frac{l^2}{(e^2 - 1)^2}$$

and

$$b^2 = \frac{l^2}{e^2 - 1}. \qquad (5.41)$$

The parameters a and b correspond to the semi-major axis and the semi-minor axis of the hyperbola, respectively (see Figure 5.8).

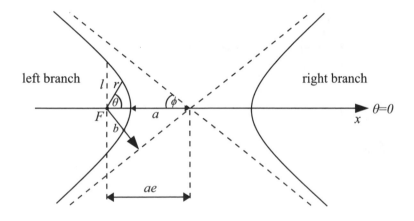

FIGURE 5.8
The hyperbolic orbit

In terms of the semi-major axis, the polar form of (5.40) is, by (5.36),

$$r = \frac{a(e^2 - 1)}{1 + e \cos \theta}. \qquad (5.42)$$

The dotted lines in Figure 5.8 are asymptotes. If ϕ is the angle subtended by an asymptote in the left branch as shown, then as $r \to \infty$, $\theta \to \pm(\pi - \phi)$, where $\cos \phi = 1/e$ and $0 < \phi < \pi/2$. The left branch of the hyperbola indicates a particle in an attractive orbit with the centre of attraction at the focus F. We will see when discussing Rutherford scattering that the right branch of the hyperbola indicates a particle in a repulsive orbit with the centre of repulsion again at the focus F. In this case $\theta \to \pm\phi$, where $\cos\phi = -1/e$ and $\phi > \pi/2$. Thus for the attractive case, a particle must orbit along the left branch.

Parabolic orbits $e = 1$

The orbital equation takes on the rather simple form

$$r = \frac{l}{1 + \cos\theta},$$ (5.43)

which describes the polar form of a parabola as depicted in Figure 5.9.

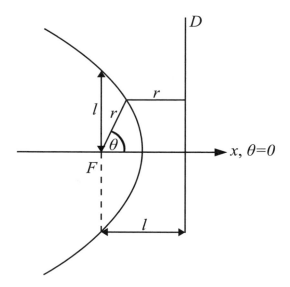

FIGURE 5.9
The parabolic orbit

The locus of the points that are equidistant from the fixed focus F and the vertical straight line D is the *directrix*. Here, the semi-latus rectum l is the distance between F and D. It is straightforward to show, from a geometrical argument, that Figure 5.9 yields (5.43) (see Appendix).

5.5.2   Orbital energy

There exists a rather simple equation relating the total energy E of a particle in orbit and the eccentricity e of the orbital path. To begin with, consider the radial energy equation (5.22). Using the chain rule on $\dot{r}$ gives

$$\dot{r} = \frac{dr}{d\theta}\dot{\theta}.$$

Substituting this into (5.22) yields

$$E = \frac{1}{2}m\left(\frac{dr}{d\theta}\right)^2 \dot{\theta}^2 + \frac{L^2}{2mr^2} + V(r).$$

As we are dealing with a central inverse square law of force, the potential energy $V(r)$ can be replaced by $-k/r$ (see (5.26)). Thus

$$E = \frac{1}{2}m\left(\frac{dr}{d\theta}\right)^2 \dot{\theta}^2 + \frac{L^2}{2mr^2} - \frac{k}{r}.$$

Replacing $\dot{\theta}$ with L/mr^2 (see (5.20)) and differentiating (5.36) with respect to θ,

$$\frac{dr}{d\theta} = \frac{le\sin\theta}{(1 + e\cos\theta)^2}$$

yields, after some algebra,

$$
\begin{aligned}
E &= \frac{L^2}{2ml^2}e^2\sin^2\theta + \frac{L^2}{2ml^2}(1 + e\cos\theta)^2 - \frac{L^2}{ml^2}(1 + e\cos\theta) \\
&= \frac{L^2}{2ml^2}(e^2 - 1) \\
&\equiv \frac{mk^2}{2L^2}(e^2 - 1). \tag{5.44}
\end{aligned}
$$

Hence, from the orbital energy equation (5.44), a particle's path will correspond to an ellipse if $E < 0$; the special case of a circle if $E = -L^2/2ml^2 \equiv -mk^2/2L^2$ (this is E_1 in Figure 5.6 — the minimum effective potential energy); a hyperbola if $E > 0$; and a parabola if $E = 0$.

The results above can be summarised in Table 5.1 (see also Figure 5.10 for a schematic representation).

TABLE 5.1
Classification of the orbits

Orbit	Class	Energy	Eccentricity
ellipse	bounded	$E < 0$	$0 < e < 1$
circle	bounded	$E = -L^2/2ml^2$	$e = 0$
hyperbola	unbounded	$E > 0$	$e > 1$
parabola	unbounded	$E = 0$	$e = 1$

The total energy of the particle in an elliptic or hyperbolic orbit is inversely proportional to the semi-major axis a. This can be seen by writing (5.44) as

$$e^2 - 1 = \frac{2EL^2}{mk^2}$$

and substituting this into (5.38) and (5.41) yielding

$$|E| = \frac{k}{2a} \quad \text{for the ellipse}$$

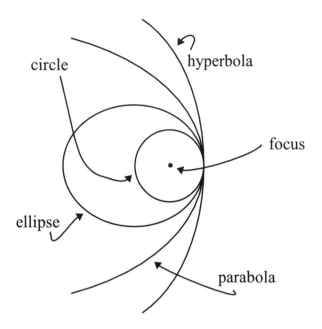

FIGURE 5.10
Schematic of the orbital paths. All orbits have been chosen to have the same
value for the closest approach

and

$$E = \frac{k}{2a} \quad \text{for the hyperbola.}$$

Similar relations for the angular momentum also follow:

$$L^2 = 2m|E|b^2 \quad \text{for the ellipse}$$

and

$$L^2 = 2mEb^2 \quad \text{for the hyperbola,}$$

where b is the semi-minor axis. Notice also that in both cases, we have $b^2 = al$.

Example 5.1 A particle of mass m orbits a planet of mass M. At
the apse $\theta = 0$ and the speed of the particle is ν, show that the total
energy of the particle is

$$E = \frac{GMm(e^2 - 1)}{2l},$$

and that the particle will escape the gravitational influence of the
planet when

$$\nu^2 \geq \frac{2GM(1 + e)}{l}.$$

Show further that if the orbit is elliptical, the speed of the particle is given by

$$v_e^2 = GM\left(\frac{2}{r} - \frac{1}{a}\right),$$

and if the orbit is parabolic, the speed of the particle is given by

$$v_p^2 = \frac{2GM}{r}.$$

Solution Using (5.22), we can write

$$E = \frac{1}{2}m\dot{r}^2 + \frac{L^2}{2mr^2} - \frac{GMm}{r}.$$

At the apse $\theta = 0$, and so $r = l/(1+e)$, the *apsidal distance*. Also, $\dot{r} = 0$ at the apse; hence,

$$E = \frac{GMm(1+e)^2}{2l} - \frac{GMm(1+e)}{l} = \frac{GMm(e^2-1)}{2l}, \qquad (5.45)$$

where $L^2 = GMm^2l$.

From conservation of energy

$$E = \frac{1}{2}mv^2 - \frac{GMm}{r}. \qquad (5.46)$$

The particle will escape the gravitational influence of the planet if $E \geq 0$. Thus

$$v^2 \geq \frac{2GM}{r} = \frac{2GM(1+e)}{l}.$$

On comparing (5.45) and (5.46), we have

$$\frac{1}{2}mv^2 - \frac{GMm}{r} = \frac{GMm(e^2-1)}{2l}$$

or

$$v^2 = GM\left(\frac{2}{r} + \frac{e^2-1}{l}\right).$$

For an elliptical orbit the semi-major axis $a = l/(1-e^2)$. Hence,

$$v_e^2 = GM\left(\frac{2}{r} - \frac{1}{a}\right).$$

For a parabolic orbit the eccentricity $e = 1$. Hence,

$$v_p^2 = \frac{2GM}{r}. \qquad \square$$

However, the same results may be obtained with the aid of (5.33) for P traversing a path $U(\theta)$ (see Exercise 5.10).

Substituting (5.48) into (5.18) with $\dot{\theta}$ being replaced by (5.20) yields

$$\ddot{\xi}(t) - \frac{L^2}{m^2(R+\xi)^3} = \frac{F(R+\xi)}{m}$$

or

$$\ddot{\xi}(t) - \frac{L^2}{m^2 R^3}(1+\xi/R)^{-3} = \frac{F(R+\xi)}{m}.$$

The second term on the left can be expanded using the binomial series:

$$(1+\xi/R)^{-3} = 1 - 3\frac{\xi}{R} + O(\xi^2).$$

The term on the right can be expanded using the Taylor series about $r = R$:

$$F(R+\xi) = F(R) + \xi F'(R) + O(\xi^2).$$

Hence,

$$\ddot{\xi}(t) - \frac{L^2}{m^2 R^3}[1 - 3(\xi/R) + O(\xi^2)] = \frac{1}{m}[F(R) + \xi F'(R) + O(\xi^2)].$$

Ignoring $O(\xi^2)$, this equation simplifies, with the aid of (5.47), to

$$\ddot{\xi}(t) - \frac{1}{m}\left(3\frac{F(R)}{R} + F'(R)\right)\xi = 0. \tag{5.49}$$

Now, we consider two cases. For

$$3\frac{F(R)}{R} + F'(R) < 0, \tag{5.50}$$

equation (5.49) is simple harmonic having trigonometric solutions: sine and cosine. As these solutions remain bounded over increasing time, the orbit described by P will be *stable* for small perturbations. If

$$3\frac{F(R)}{R} + F'(R) > 0,$$

equation (5.49) will yield exponential solutions, one of which will not remain bounded over increasing time. The orbit described by P will be *unstable* for small perturbations.

In particular, consider the case of a force governed by a power law

$$F(r) = Kr^n.$$

If we are interested only in forces that result in stable circular orbits, then from (5.50), we must have the following relation satisfied for n:

$$n > -3.$$

Thus, for the linearised theory, both Hooke's law (direct distance law), $n = 1$, and the inverse square law, $n = -2$, give stable circular orbits; however, the inverse cube law, $n = 3$, gives unstable circular orbits. The higher order terms that were neglected in our analysis above *may* become significant as the number of orbits increases. Thankfully, a remarkable result known as *Bertrand's theorem*[6] concurs with the situations above for the linearised theory. It states: *The only central forces that result in particle orbits being stable and closed are Hooke's law and the inverse square law.*

5.7 Kepler's laws of planetary motion

It was mentioned in Section 1.3 that through the work of Tycho Brahe and Johannes Kepler, Newton formulated his law of gravitation: the (attractive) inverse square law of gravitation.

Tycho Brahe gathered enormous amounts of data of stunning accuracy for the period, through observations of the known planets in the solar system of the 16th century. It was left to Johannes Kepler (Tycho Brahe's pupil) to analyse the accumulated information, which took over 20 years to complete. Kepler then deduced his now legendary three laws of planetary motion:

(KI)   *Kepler's first law*. The planets move in ellipses with the Sun at one focus.

(KII)   *Kepler's second law*. The line joining any given planet and the Sun sweeps out equal areas in equal times.

(KIII)   *Kepler's third law*. The square of the period of the orbit is proportional to the cube of the semi-major axis.

Kepler's first law follows from our discussions on orbit theory; namely, that planets move in closed orbits and these closed orbits are elliptical with the Sun at one focus being the centre of force. Kepler's second law can be obtained as follows.

The area δA swept out by the radius vector (joining the origin to the planet) when the planet moves through an angle $\delta\theta$ in time δt is (see Figure 5.11)

$$\delta A = \frac{1}{2} r^2 \delta\theta$$

giving the rate of change of area

$$\frac{dA}{dt} = \frac{1}{2} r^2 \dot\theta = \frac{L}{2m}, \tag{5.51}$$

---

[6] Joseph Louis François Bertrand (1822-1900) discovered this law in 1873.

5.9.1 Scattering angle and impact parameter

The force F between a particle of charge q and a particle of charge Q is given by Coulomb's law:

$$F = \frac{qQ}{4\pi\epsilon_0 r^2}, \tag{5.56}$$

where ϵ_0 is a constant called the *permittivity of free space*. If both charges are positive (or negative) then $qQ > 0$ and the force between them is repulsive. Indeed, Coulomb's law is a repulsive inverse square law of force.

One can achieve the same result by letting $k = -qQ/4\pi\epsilon_0$ in (5.27). The corresponding potential energy is then, by (5.26),

$$V(r) = \frac{qQ}{4\pi\epsilon_0 r}, \tag{5.57}$$

which from (5.28) gives a total radial energy

$$E = \frac{1}{2}m\dot{r}^2 + \frac{qQ}{4\pi\epsilon_0 r} + \frac{L^2}{2mr^2}. \tag{5.58}$$

It was already established in Section 5.4 that a particle travelling from infinity ($V(r) = 0$) will eventually reach a minimum radial distance from an origin before travelling out towards infinity again. The path of the particle is necessarily hyperbolic.

Now, consider the particle with charge Q fixed at an origin O and the particle with charge q approaching the origin from infinity with speed v — see Figure 5.13 and compare with Figure 5.8. Because the particle with charge q approaches from infinity, the potential energy $V(r)$ (and the effective potential energy) will be zero. So the total energy, from (5.58), will be purely kinetic:

$$E = \frac{1}{2}mv^2. \tag{5.59}$$

The particle then moves along the hyperbola as indicated and reaches a minimum distance (closest approach) d. The angle through which the particle is deflected is Θ, known as the *scattering angle*. The particle then continues on to infinity. Had the particle with charge Q at O been absent, our travelling particle would have passed O at a distance b; this distance is called the *impact parameter*.

Angular momentum is given by $\mathbf{L} = m\mathbf{r} \times \dot{\mathbf{r}}$. When the particle is at a distance b from O, $\mathbf{v}$ is perpendicular to $\mathbf{r}$. Thus

$$|\mathbf{L}| \equiv L = mbv. \tag{5.60}$$

On substituting (5.59) and (5.60) into (5.58), we have

$$\frac{1}{2}mv^2 = \frac{qQ}{4\pi\epsilon_0 d} + \frac{m^2v^2b^2}{2md^2}. \tag{5.61}$$

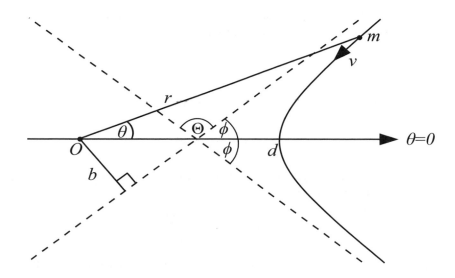

FIGURE 5.13
Particle of mass m and charge q moving along a hyperbolic path and being deflected through an angle Θ by a particle of charge Q at O

(Note that $\dot{r} = 0$ when $r = d$.) Rearranging then gives

$$d^2 - Kd - b^2 = 0,$$

where $K = qQ/4\pi\epsilon_0 mv^2$. Thus, the minimum distance d in terms of the impact parameter is

$$d = K + \sqrt{K^2 + b^2}.$$

It is also possible to find an equation that relates the scattering angle Θ to the impact parameter: this leads to a rather startling result.

From (5.41), we can write the impact parameter in terms of the semi-major axis a and the eccentricity e for the hyperbolic path of a particle. Thus

$$b^2 = a^2(e^2 - 1). \tag{5.62}$$

Furthermore, it was shown for the repulsive case that $\cos\phi = 1/e$ and from Figure 5.13

$$\Theta = \pi - 2\phi.$$

Hence, in terms of the eccentricity

$$\Theta = \pi - 2\cos^{-1}(1/e)$$

or

$$e = \sec[(\pi - \Theta)/2]. \tag{5.63}$$

Recall that for a hyperbola $a = k/2E$, and on incorporating our initial energy conditions, $E = 1/2mv^2$, we have

$$a = \frac{k}{mv^2}. \tag{5.64}$$

Substituting (5.63) and (5.64) into (5.62) yields

$$b = \frac{|k|}{mv^2} \cot(\Theta/2),$$

where the familiar trigonometric identity for $\sec^2$ has been used. Rearranging this equation finally gives

$$\Theta = 2\tan^{-1}\left(\frac{|k|}{bv^2}\right).$$

What is remarkable about this formula? Well, an α-particle fired directly at the gold nucleus would imply that $b = 0$ and the α-particle would be deflected by π; this result is independent of the initial speed v of the particle. Rutherford was quoted as saying: "it was quite the most incredible event that has ever happened to me in my life. It was almost as incredible as if you fired a 15-inch shell at a piece of tissue paper and it came back and hit you."[9]

5.10   Exercises

1. Using Kepler's third law, determine the height above the Earth's surface of a satellite performing a circular orbit and having a period of 0.07 days.

    [Use the following information to aid the calculation:

    radius of Earth = 6371 km

    distance of Moon from Earth's centre = 384000 km

    period of Moon = 28 days.]

2. Show that the period of a satellite that orbits close to the surface of a sphere depends on the mean density of the sphere ρ and the universal constant G.

    Find the period of a satellite that orbits close to the surface of Jupiter, given that the mean density of Jupiter is 1326 kgm^{-3} and $G = 6.67 \times 10^{-11}$ Nm^{2}kg^{-2}.

---

[9] Andrade, E.N. da C, *Rutherford and the Nature of the Atom*; Anchor Books: Garden City, N.Y.: 1964; p.111.

3. The period of a satellite orbiting the planet Mars (mass 6.418×10^{23} kg) is 1.1382 days. Taking $G = 6.67 \times 10^{-11}$ Nm^{2}kg^{-2}, calculate the semi-major axis of the satellite's orbit. If the perigee and apogee of the orbiting satellite is 1500 km and 35800 km, respectively, estimate the average value of Mars' diameter.

4. Consider a satellite moving under the influence of a central force $d_1 r^{-2} + d_2 r^{-4}$, where d_1 and d_2 are constants. Show that the orbit is stable if and only if $r_0^2 > \frac{d_2}{d_1}$, where r_0 is the radius of the circular orbit.

   If the force of attraction is now given by $m d_3 r^{-2} e^{-r^2}$, where d_3 is a constant and m is the mass of the satellite, show that stability occurs when $r_0^2 > \frac{1}{2}$.

5. Show that for a particle following a hyperbolic orbit, the speed of the particle ν_{h} can be related by

$$\nu_{\mathrm{h}}^2 = GM\left(\frac{2}{r} + \frac{1}{a}\right),$$

   where M is the mass of the source of the gravitational force and a is the semi-major axis of the orbital path.

6. The position of a particle moving in the xy-plane is given by

$$\mathbf{r}(t) = (a\cos\omega t, b\sin\omega t), \qquad a > b,$$

   where a, b and ω are positive constants. Show that the path of the particle is elliptic.

7. Consider a particle P of mass m initially perched on top of a smooth sphere of radius R and centre O (see Figure 5.14). The particle is then given a small horizontal nudge so that it begins to slide down the surface of the sphere without rolling. The normal reaction of the sphere is N. Find the position and angular speed of P as it leaves the surface of the sphere. [Hint: What is the value of N as P leaves the sphere?]

8. A particle of mass m describes a closed orbit under a force k/r^2 about the Sun S. Show that the polar forms of the particle's radial and transverse components of acceleration are given, respectively, by

$$\ddot{r} - r\dot{\theta}^2 \qquad \text{and} \qquad \frac{1}{r}\frac{d}{dt}(r^2\dot{\theta}).$$

   The particle passes through the point P_0 with speed v_0 such that the angle between the velocity vector and SP_0 is α. Using the equations of motion, show that

$$\dot{\theta} = \frac{v_0 r_0 \sin\alpha}{r^2},$$

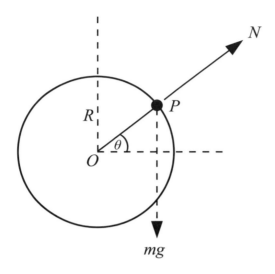

FIGURE 5.14
Particle on a smooth sphere

where r_0 is the distance of P_0 from the origin. Hence, multiply the radial equation of motion by $\dot{r}$ and integrate to show that

$$v^2 - v_0^2 = \frac{2k(r_0 - r)}{mr_0 r},$$

where k is a constant. Deduce the equation of conservation of energy and show that the orbit is elliptical provided that

$$v_0^2 < \frac{2k}{mr_0}.$$

9. A particle of mass m moves under the influence of an attractive force

$$\mathbf{F} = -\frac{m\beta}{r^4}\hat{\mathbf{r}},$$

where β is a positive constant. If the particle traverses a circular path of radius R, show that the angular speed ω is

$$\omega = \frac{\sqrt{\beta}}{R^3}$$

and the period τ is

$$\tau = \frac{2\pi R^3}{\sqrt{\beta}}.$$

Now suppose the particle traverses a new path $r = R(1 + \xi(t))$, where $\xi(t)$ is a small perturbation. Use the radial equation and a Taylor expansion to show that

$$\ddot{\xi}(t) = \frac{2\beta}{R^6}\xi,$$

and justify that the orbit is unstable.

10. A particle P moves under the influence of an attractive central force, $F(r) < 0$, such that the orbit is circular $r = R$. Now let P be perturbed by a small amount $\xi(\theta)$ so that the path of P is given by

$$u = [1 + \xi(\theta)]/R.$$

Use the orbital equation (5.33) and equation (5.47) to show that

$$\xi''(\theta) + \left(3 + R\frac{F'(R)}{F(R)}\right)\xi = 0,$$

where the prime denotes differentiation with respect to θ. If the orbit is now stable, show that the orbit will be closed provided that

$$3 + R\frac{F'(R)}{F(R)} \in \mathbb{Z}^+,$$

and deduce that

$$F(r) = Kr^n,$$

where $n > -3$.

11. Suppose there exists a universe in which the gravitational fields of all massive objects are inverse-cubed in nature. Consider a particle of mass m subject to a central inverse cube law of attraction

$$\mathbf{F} = -\frac{m\beta}{r^3}\hat{\mathbf{r}}. \tag{5.65}$$

Derive the equation of motion of the particle in this field and show that the orbital equation is

$$\frac{d^2u}{d\theta^2} - \Omega^2 u = 0,$$

where Ω^2 needs to be found. By considering cases $\Omega^2 > 0$, $\Omega^2 = 0$ and $\Omega^2 < 0$, show that the only bounded orbit is circular. Would such an orbit remain stable?

12. A particle of mass m approaches a planet that exerts a force given by (5.65) in Exercise 5.11 with $\beta = 3L^2/4m^2$. Use the orbital equation of Exercise 5.11 together with the conditions $u(0) = 0$ and $u'(0) =$

$1/b$ (where b is the distance between the particle's uninterrupted path and the planet had the planet not been there) to obtain the path of the particle. Use conservation of angular momentum to find the speed of the particle at perigee if the particle's initial speed of approach was W.

13. Integrate (5.20) over the limits $[0, \tau]$ and use (5.36) to obtain Kepler's third law.

14. A beam of particles is fired at a fixed target nucleus. Each particle has unit mass and is travelling at speed v. The nucleus exerts a repulsive force

$$\mathbf{F} = \frac{\beta^2}{r^3}\hat{\mathbf{r}},$$

where β is a positive constant. Show that the impact parameter b satisfies

$$\frac{bv}{\beta} = (\pi - \Theta)[(2\pi - \Theta)\Theta]^{-\frac{1}{2}},$$

where Θ is the scattering angle.

6

Multi-Particle Systems

CONTENTS

Our discussion in previous chapters concentrated primarily on single-particle motion.[1] Even when two separate masses appeared in a single system, such as a satellite orbiting the Earth, the Earth was considered fixed relative to the orbiting satellite, thus only the motion of single particles was effectively examined. Of course, a satellite is composed of many particles, each exerting a force on the others. However, our treatment of orbits assumed that the total mass of the satellite is concentrated at the *centre of mass* point. Under these circumstances, our analysis of the dynamical behaviour of the satellite reduced to an analysis of the dynamical behaviour of the centre of mass (CM) point subject to external forces. Indeed, our discussion of multi-particle systems will highlight the fact that the motion of individual particles within the system need not be analysed if our interest lies only in how the system behaves as a whole.

---

[1]Recall that we briefly encountered a two-particle system in Section 1.2.

6.1   Conservation of linear momentum

Let our multi-particle system be comprised of N particles. Each particle is labelled by an index i where $i = 1, 2, \ldots, N$. Thus, particle 1 has a mass m_1, particle 2 has mass m_2 and the ith particle has mass m_i. At some instant in time these N particles are located at points having position vectors $\mathbf{r}_1$, $\mathbf{r}_2$ and $\mathbf{r}_i$, respectively. So the linear momentum of the ith particle is $\mathbf{P}_i = m_i \dot{\mathbf{r}}_i$.

Now, for any two particles labelled i and j the force acting on the ith particle exerted by the jth particle is $\mathbf{F}_{ij}$. Forces labelled with a double index in this manner are *internal forces* — they arise only within the system. Also, particles within the system may be subject to *external forces* — they arise outside of the system and are independent of the system, such as an external gravitational field. An external force acting on the ith particle is denoted by $\mathbf{F}_i^{\text{ext}}$.

From Newton's second law of motion, we can write the equation of motion for each particle in the system:

$$\dot{\mathbf{P}}_i = m_i \ddot{\mathbf{r}}_i = \mathbf{F}_i^{\text{ext}} + F_{i1} + F_{i2} + \ldots + F_{iN} = \mathbf{F}_i^{\text{ext}} + \sum_{j \neq i}^{N} \mathbf{F}_{ij}. \tag{6.1}$$

According to Newton's third law of motion, $\mathbf{F}_{ij} = -\mathbf{F}_{ji}$. As the ith particle cannot exert a force upon itself, we have $\mathbf{F}_{ii} = \mathbf{0}$ and the sum in (6.1) extends over $(N - 1)$ particles.

Let the *total mass* of the system be defined as

$$M = \sum_{i=1}^{N} m_i, \tag{6.2}$$

then the *position of the centre of mass* is

$$\mathbf{R} = \frac{m_1 \mathbf{r}_1 + m_2 \mathbf{r}_2 + \ldots + m_N \mathbf{r}_N}{m_1 + m_2 + \ldots + m_N} = \frac{1}{M} \sum_{i=1}^{N} m_i \mathbf{r}_i. \tag{6.3}$$

It is now convenient to define the *total linear momentum* $\mathbf{P}$ of the system as the sum of the individual momenta. Thus,

$$\mathbf{P} = \sum_{i=1}^{N} \mathbf{P}_i = \sum_{i=1}^{N} m_i \dot{\mathbf{r}}_i = M \dot{\mathbf{R}}. \tag{6.4}$$

This equation is telling us that if the total mass of the system were concentrated into a single particle of mass M located at the centre of mass of the system, the total linear momentum of the system would be equivalent to the linear momentum of the particle of mass M.

The way in which the centre of mass moves can be ascertained by appealing to (6.1). Thus,

$$\dot{\mathbf{P}} = \sum_{i=1}^{N} \dot{\mathbf{P}}_i = \sum_{i=1}^{N} \left(\mathbf{F}_i^{\text{ext}} + \sum_{j \neq i} \mathbf{F}_{ij} \right)$$

$$= \sum_{i=1}^{N} \mathbf{F}_i^{\text{ext}} + \sum_{i=1}^{N} \sum_{j \neq i}^{N} \mathbf{F}_{ij}$$

$$= \sum_{i=1}^{N} \mathbf{F}_i^{\text{ext}} + \sum_{i=1<j}^{N} \left(\mathbf{F}_{ij} + \mathbf{F}_{ji} \right). \tag{6.5}$$

Notice that the double sum occurring in the second line has a total of $N(N-1)$ terms — the terms $\mathbf{F}_{11}, \mathbf{F}_{22}, \ldots, \mathbf{F}_{NN}$ are excluded due to the argument above. However, on summing, it can be seen that whenever the term $\mathbf{F}_{ij}$ appears, the corresponding term $\mathbf{F}_{ji}$ arises: $\mathbf{F}_{12} + \mathbf{F}_{21}, \mathbf{F}_{13} + \mathbf{F}_{31}$, etc. Hence, the third line follows directly from the second line where the restriction $i = 1 < j$ ensures that the value of the index is not repeated in the same term, and for any value i, the value of j is strictly greater in that term.

Newton's third law of motion $\mathbf{F}_{ij} = -\mathbf{F}_{ji}$ can now be applied in a very obvious way, so that the second sum on the right in the third line vanishes. Therefore, the rate of change of linear momentum for a multi-particle system is equal only to the net *external* force acting on the system

$$\dot{\mathbf{P}} = \sum_{i=1}^{N} \mathbf{F}_i^{\text{ext}} \equiv \mathbf{F}^{\text{ext}}. \tag{6.6}$$

If no external forces act on the system it is evident that

$$\dot{\mathbf{P}} = \mathbf{0}, \tag{6.7}$$

which implies

$$\mathbf{P} = \text{constant}. \tag{6.8}$$

Equation (6.8) is the familiar law of conservation of linear momentum only for a multi-particle system.

The remarkable conclusion arising from the discussion above is that for a system of N particles, one may replace the total mass of the system with a single particle of mass M located at the centre of mass of the system. Thus the dynamical behaviour of the system (excluding internal motion such as spinning) can be modelled by the motion of the centre of mass particle through space. And it is precisely that which allowed us to describe motion of planets and satellites in their various orbits. There is, however, one caveat that must be observed: *in general*, the external forces acting on a system of particles is not equal to the same forces acting on the centre of mass particle. The way in which the individual particles are distributed throughout the system will have an effect, as would a non-uniform gravitational field. Furthermore, if our system happened to be a kicked rugby ball[2] and our interest lay in what part of the ball first came into contact with the ground, then knowledge of the position of the centre of mass would be irrelevant.

6.2   Conservation of angular momentum

The total angular momentum of the N-particle system about the origin is defined as[3]

$$\mathbf{L} = \sum_{i=1}^{N} \mathbf{r}_i \times \mathbf{P}_i. \tag{6.9}$$

It follows that the rate of change of angular momentum is

$$\dot{\mathbf{L}} = \sum_{i=1}^{N} \left(\dot{\mathbf{r}} \times \mathbf{P}_i + \mathbf{r}_i \times \dot{\mathbf{P}}_i \right) = \sum_{i=1}^{N} \mathbf{r}_i \times \dot{\mathbf{P}}_i.$$

(Note that $\dot{\mathbf{r}}_i$ and $\mathbf{P}_i$ are parallel vectors.) This can be written with the aid of (6.1) as

$$
\begin{aligned}
\dot{\mathbf{L}} &= \sum_{i=1}^{N} \mathbf{r}_i \times \left(\mathbf{F}_i^{\text{ext}} + \sum_{j \neq i}^{N} \mathbf{F}_{ij} \right) \\
&= \sum_{i=1}^{N} \mathbf{r}_i \times \mathbf{F}_i^{\text{ext}} + \sum_{i=1}^{N} \sum_{j \neq i}^{N} \mathbf{r}_i \times \mathbf{F}_{ij} \\
&= \sum_{i=1}^{N} \mathbf{r}_i \times \mathbf{F}_i^{\text{ext}} + \sum_{i=1<j}^{N} \left(\mathbf{r}_i \times \mathbf{F}_{ij} + \mathbf{r}_j \times \mathbf{F}_{ji} \right).
\end{aligned} \tag{6.10}
$$

Note that in going from the second line to the third line in (6.10), we have adopted the same procedure that allowed us to go from line two to line three in (6.5). (One can easily justify that the second and third lines are equivalent by writing out a handful of terms in each case.)

By employing Newton's third law $\mathbf{F}_{ij} = -\mathbf{F}_{ji}$ equation (6.10) can be written as

$$\dot{\mathbf{L}} = \mathbf{G} + \sum_{i=1<j}^{N} (\mathbf{r}_i - \mathbf{r}_j) \times \mathbf{F}_{ij},$$

---

[2] A football in the USA.

[3] We could have adopted the convention where $\mathbf{L}_o$ is used to represent the angular momentum about the origin. However, if the angular momentum (or other appropriate quantity) is taken about a fixed point other than the origin, it will be denoted as such with a corresponding index.

where $\mathbf{G} = \sum_{i=1}^{N} \mathbf{r}_i \times \mathbf{F}_i^{\text{ext}}$ is called the *total external torque*.

Although not true in general[4] it is certainly true for the gravitational and electric cases that all forces are central forces. Thus, the vector $(\mathbf{r}_i - \mathbf{r}_j)$ joining two particles P_i and P_j (see Figure 6.1) is parallel to the force $\mathbf{F}_{ij}$. This implies that $(\mathbf{r}_i - \mathbf{r}_j) \times \mathbf{F}_{ij} = \mathbf{0}$. This extra condition in combination with $\mathbf{F}_{ij} = -\mathbf{F}_{ji}$ is sometimes referred to as the *strong form of Newton's third law of motion*.

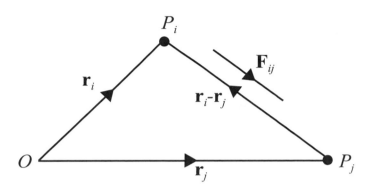

FIGURE 6.1
$\mathbf{F}_{ij}$ is a central force so $(\mathbf{r}_i - \mathbf{r}_j) \times \mathbf{F}_{ij} = \mathbf{0}$

Finally, one arrives at

$$\dot{\mathbf{L}} = \mathbf{G}, \tag{6.11}$$

and if the total external torque is zero, the system's angular momentum is constant

$$\mathbf{L} = \text{constant}, \tag{6.12}$$

which is the law of conservation of angular momentum for a multi-particle system.

6.3 The centre of mass frame

There are certain situations where it is more convenient to discuss the motion of a multi-particle system with respect to a reference frame that has the centre of mass as the origin. This frame is known as *the centre of mass frame* (see Figure 6.2).

In Figure 6.2, $\mathbf{R}$ is the position of the centre of mass and $\mathbf{r}_i^c$ is the position

---

[4]For example, the magnetic forces between moving charged particles are not central forces.

So the total angular momentum of the multi-particle system about the origin O is a combination of angular momentum $(\mathbf{R} \times \mathbf{P})$ about the origin O, where the total mass is concentrated at the centre of mass, and the angular momentum $(\mathbf{L}^c)$ of the system about the centre of mass.

6.3.3 Torque

We wish to show that the rate of change of angular momentum about the centre of mass is equal to the total external torque about the centre of mass.

Differentiating (6.17) yields

$$\dot{\mathbf{L}}^{c} = \sum_{i=1}^{N} \dot{\mathbf{r}}_i^c \times (m_i \dot{\mathbf{r}}_i^c) + \sum_{i=1}^{N} \mathbf{r}_i^c \times (m_i \ddot{\mathbf{r}}_i^c).$$

The first sum clearly vanishes. Using (6.13), we can replace $\ddot{\mathbf{r}}_i^c$ by $\ddot{\mathbf{r}}_i - \ddot{\mathbf{R}}$. On expanding the cross product this gives

$$\dot{\mathbf{L}}^{c} = \sum_{i=1}^{N} \mathbf{r}_i^c \times (m_i \ddot{\mathbf{r}}_i) - \left(\sum_{i=1}^{N} m_i \mathbf{r}_i^c \right) \times \ddot{\mathbf{R}}.$$

The second sum in this expression vanishes due to (6.14). This expression, with the aid of (6.1), can be written as

$$\dot{\mathbf{L}}^{c} = \sum_{i=1}^{N} \mathbf{r}_i^c \times \dot{\mathbf{P}}_i = \sum_{i=1}^{N} \mathbf{r}_i^c \times \left(\mathbf{F}_i^{\text{ext}} + \sum_{j \neq i}^{N} \mathbf{F}_{ij} \right)$$

$$= \sum_{i=1}^{N} \mathbf{r}_i^c \times \mathbf{F}_i^{\text{ext}} + \sum_{i=1<j}^{N} (\mathbf{r}_i^c - \mathbf{r}_j^c) \times \mathbf{F}_{ij}$$

$$= \sum_{i=1}^{N} \mathbf{r}_i^c \times \mathbf{F}_i^{\text{ext}}, \tag{6.18}$$

where the third line arises on appealing to the strong form of Newton's third law of motion. As the final term is the total external torque about the centre of mass, we finally arrive at

$$\dot{\mathbf{L}}^{c} = \mathbf{G}^c. \tag{6.19}$$

A number of conclusions can be drawn from (6.19). In particular, the equation holds true if the centre of mass is accelerating (and, therefore, not fixed in any inertial frame) due to a net external force acting on the system. If the total external torque vanishes, the angular momentum remains constant, regardless of how the particles within the system move. For example, a high-diver can reorientate his body in mid-air so that he enters the water head first; however, during the process of falling, he is unable to affect the trajectory of his centre of mass.[5] This is also true if a multi-particle system were subject to a uniform gravitational field. The angular momentum relative to the centre of mass is constant because gravity acts at the centre of mass.

6.3.4 A caveat to the torque equation

In deriving (6.19), we have applied one unnecessary condition: the strong form of Newton's third law of motion.

It is true that in the case of gravity applying this law is perfectly legitimate; however, equation (6.19) should be deduced even if the internal forces do not act centrally. Under these circumstances a *couple*[6] will arise (see Figure 6.3). Nevertheless, we still see that ultimately, all torques are neutralised by other internal forces of the system.

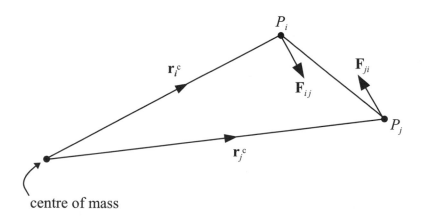

FIGURE 6.3
Two forces producing a torque that constitutes a couple

Now, if $\mathbf{F}$ is a conservative force (not necessarily central) we can write

$$\mathbf{F} = -\boldsymbol{\nabla}V \equiv -\frac{\partial}{\partial r_i}V(r_i), \quad i = 1, 2, 3 \quad r_i = (x, y, z),$$

where V is a single valued function called the potential energy (see Section 4.2 for further clarification). The work done in moving a particle from one point to another is independent of the path joining the two points.

Similarly, the internal torque of a system about some axis n, say, denoted

---

[5]Unless, of course, the said diver happens to reside in a Marvel comic!
[6]A pair of local forces equal in magnitude and opposite in direction constitute a couple, provided that the forces do not act along the same line.

by G_n, can be related to the potential energy $V(\theta)$ by

$$G_n = \frac{-dV(\theta)}{d\theta},$$

where θ is the angle through which the system will rotate. However, our system is rotationally invariant; that is, V is independent of the system's orientation. Hence, $G_n = 0$ and again we arrive at equation (6.19).

6.3.5   Energy

The total kinetic energy of a multi-particle system is defined as

$$T = \sum_{i=1}^{N} \frac{1}{2} m_i \dot{\mathbf{r}}_i^2. \tag{6.20}$$

Substituting (6.13) into this yields

$$T = \sum_{i=1}^{N} \frac{1}{2} m_i \dot{\mathbf{R}}^2 + \sum_{i=1}^{N} \frac{1}{2} m_i (\dot{\mathbf{r}}_i^c)^2 + \left(\sum_{i=1}^{N} m_i \dot{\mathbf{r}}_i^c \right) \cdot \dot{\mathbf{R}}$$

$$= \sum_{i=1}^{N} \frac{1}{2} m_i \dot{\mathbf{R}}^2 + T^c, \tag{6.21}$$

where

$$T^c = \sum_{i=1}^{N} \frac{1}{2} m_i (\dot{\mathbf{r}}_i^c)^2 \tag{6.22}$$

is the sum of the kinetic energies of the individual particles relative to the centre of mass.

To establish the total energy in the centre of mass frame, we first need to differentiate (6.20) with respect to time.

$$\dot{T} = \sum_{i=1}^{N} \dot{\mathbf{r}}_i \cdot (m_i \ddot{\mathbf{r}}_i)$$

$$= \sum_{i=1}^{N} \dot{\mathbf{r}}_i \cdot \left(\mathbf{F}_i^{\text{ext}} + \sum_{j \neq i}^{N} \mathbf{F}_{ij} \right)$$

$$= \sum_{i=1}^{N} \dot{\mathbf{r}} \cdot \mathbf{F}_i^{\text{ext}} + \sum_{i=1}^{N} \sum_{j \neq i}^{N} \dot{\mathbf{r}}_i \cdot \mathbf{F}_{ij}$$

$$= \sum_{i=1}^{N} \dot{\mathbf{r}}_i \cdot \mathbf{F}_i^{\text{ext}} + \sum_{i=1<j}^{N} (\dot{\mathbf{r}}_i - \dot{\mathbf{r}}_j) \cdot \mathbf{F}_{ij}, \tag{6.23}$$

where we have used Newton's third law to obtain the last line.

The last two terms in (6.23) represent the rate at which work is done by the external forces and the rate at which work is done by the internal forces, respectively. There is no general justification for assuming that the first of these terms will vanish. This is because at any instant the position of the particle may change even though the potential energy is a function of position. (However, when we come to discuss *rigid body motion*, the particles will maintain a fixed separation with respect to each other. In this case, the forces between constituent particles are *central* and no work is done by the forces.) Assuming that the internal and external forces are *conservative fields*, they can be written as gradients of the internal and external potential energies:

$$\mathbf{F}_{ij} = -\boldsymbol{\nabla}_i V_{ij}(|\mathbf{r}_i - \mathbf{r}_j|) \tag{6.24}$$

and

$$\mathbf{F}_i^{\text{ext}} = -\boldsymbol{\nabla}_i V_i(\mathbf{r}_i), \tag{6.25}$$

respectively. The *internal potential energy* for a pair of particles i and j is represented by V_{ij}, whilst the external potential energy due to an external force acting on the ith particle is represented by V_i. The *grad* operator is defined as $\boldsymbol{\nabla}_i = \partial/\partial \mathbf{r}_i$. Notice that the internal energy for a pair of particles depends on the distance between the particles, whereas the external potential energy is coordinate dependent; more about this presently.

Denoting the internal potential energy V^{int} as the sum $\sum_{i=1<j}^N V_{ij}$ of all particle pairs, which number $N(N-1)/2$ in total, the rate at which work is done by the internal forces is $-V^{\text{int}}$. As this represents the last term in (6.23), we can write

$$\dot{T} + \dot{V}^{\text{int}} = \sum_{i=1}^N \dot{\mathbf{r}}_i \cdot \mathbf{F}_i^{\text{ext}}, \tag{6.26}$$

where assumption (6.25) is yet to be employed. Substituting (6.21) into this equation and appealing to (6.13) yields

$$M\dot{\mathbf{R}} \cdot \ddot{\mathbf{R}} + \dot{T}^{\text{c}} + \dot{V}^{\text{int}} = \dot{\mathbf{R}} \cdot \sum_{i=1}^N \mathbf{F}_i^{\text{ext}} + \sum_{i=1}^N \dot{\mathbf{r}}_i^{\text{c}} \cdot \mathbf{F}_i^{\text{ext}}.$$

But

$$\dot{\mathbf{R}} \cdot (M\ddot{\mathbf{R}}) = \dot{\mathbf{R}} \cdot \dot{\mathbf{P}} = \dot{\mathbf{R}} \cdot \sum_{i=1}^N \mathbf{F}_i^{\text{ext}},$$

from (6.4) and (6.6). So, if we do *not* employ the assumption (6.25),

$$\dot{T}^{\text{c}} + \dot{V}^{\text{int}} = \sum_{i=1}^N \dot{\mathbf{r}}_i^{\text{c}} \cdot \mathbf{F}_i^{\text{ext}}. \tag{6.27}$$

Denoting the *external potential energy* V^{ext} as the sum $\sum_{i=1}^N V_i$ of all N-particles, the rate at which work is done by the external forces is $-\dot{V}^{\text{ext}}$. As this represents the summed term in (6.26), we have

$$\dot{T}^c + \dot{V}^{\text{int}} + \dot{V}^{\text{ext}} = 0.$$

Hence, the law of conservation of energy is

$$T^c + V^{\text{int}} + V^{\text{ext}} = E \tag{6.28}$$

or

$$T^c + \sum_{i=1}^{N} V_i(\mathbf{r}_i) + \sum_{i=1<j}^{N} V_{ij}(|\mathbf{r}_i - \mathbf{r}_j|) = E. \tag{6.29}$$

The conservation of energy equation (6.29) reiterates what was said above regarding the coordinate dependency of the internal potential energy and the difference-in-coordinate dependency of the external potential energy. From this, one can conclude that the internal potential energy is frame dependent, whilst the external potential energy is frame independent.

> **Example 6.1** For a multi-particle system, consider an arbitrarily chosen not necessarily inertial reference frame with origin O' moving with respect to a fixed inertial frame with origin O. Show that the rate of change of angular momentum about O' and the external torque about O' will not be related by a form similar to (6.11) nor (6.19), but will incorporate an extra term that should be determined. Comment on the result.
>
> **Solution** Let us first draw a diagram, similar to Figure 6.2 to show the position of the arbitrary reference point O', the position of the centre of mass and the position of an arbitrary particle labelled P_i (Figure 6.4).

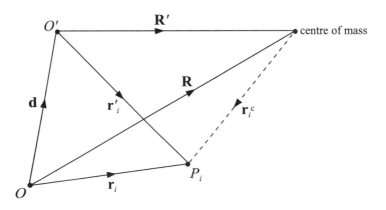

FIGURE 6.4
Schematic diagram for Example 6.1

Let the position of O' relative to O be the vector $\mathbf{d}(t)$. Thus, in terms of $\mathbf{d}$ the position of the centre of mass and the position of the ith particle are given, respectively, by

$$\mathbf{R} = \mathbf{d} + \mathbf{R}' \tag{6.30}$$

$$\mathbf{r}_i = \mathbf{d} + \mathbf{r}'_i. \tag{6.31}$$

Now, by analogy to (6.16), the angular momentum of the system about O' is

$$\mathbf{L}_{o'} = M\mathbf{R}' \times \dot{\mathbf{R}}' + \mathbf{L}^c, \tag{6.32}$$

where

$$\mathbf{R}' = \frac{1}{M} \sum_{i=1}^{N} m_i \mathbf{r}'_i,$$

and $\mathbf{L}^c$ is given by (6.17). Substituting (6.32) into (6.16) enables us to eliminate the angular momentum about the centre of mass. Thence,

$$\mathbf{L} = M\mathbf{R} \times \dot{\mathbf{R}} - M\mathbf{R}' \times \dot{\mathbf{R}}' + \mathbf{L}_{o'}.$$

Differentiating yields

$$\dot{\mathbf{L}} = \mathbf{R} \times (M\ddot{\mathbf{R}}) - \mathbf{R}' \times (M\ddot{\mathbf{R}}') + \dot{\mathbf{L}}_{o'}$$
$$= (\mathbf{d} + \mathbf{R}') \times M\ddot{\mathbf{R}} - \mathbf{R}' \times M(\ddot{\mathbf{R}} - \ddot{\mathbf{d}}) + \dot{\mathbf{L}}_{o'}$$

on employing (6.30). This simplifies to

$$\dot{\mathbf{L}} = \mathbf{d} \times \mathbf{F}^{\text{ext}} + \mathbf{R}' \times M\ddot{\mathbf{d}} + \dot{\mathbf{L}}_{o'}, \tag{6.33}$$

where $\mathbf{F}^{\text{ext}}$ is defined by (6.6).

The torque about O is

$$\begin{aligned} \mathbf{G} &= \sum_{i=1}^{N} \mathbf{r}_i \times \mathbf{F}_i^{\text{ext}} \\ &= \sum_{i=1}^{N} (\mathbf{d} + \mathbf{r}'_i) \times \mathbf{F}_i^{\text{ext}}, \quad \text{using (6.31)} \\ &= \mathbf{d} \times \sum_{i=1}^{N} \mathbf{F}_i^{\text{ext}} + \sum_{i=1}^{N} \mathbf{r}'_i \times \mathbf{F}_i^{\text{ext}} \\ &= \mathbf{d} \times \mathbf{F}^{\text{ext}} + \mathbf{G}_{o'}. \end{aligned} \tag{6.34}$$

But, with respect to O the rate of change of angular momentum is equal to the torque. Hence, equating (6.33) and (6.34) directly yields

$$\dot{\mathbf{L}}_{o'} = \mathbf{G}_{o'} - \mathbf{R}' \times M\ddot{\mathbf{d}}. \tag{6.35}$$

Thus, determining the rate of change of angular momentum for a system about a point other than the origin or centre of mass produces an extra term $-\mathbf{R}' \times M\ddot{\mathbf{d}}$. This term will vanish if the acceleration of O' vanishes ($\ddot{\mathbf{d}} = \mathbf{0}$) or if it is directed along the line joining O' to the centre of mass ($\ddot{\mathbf{d}} \uparrow\uparrow \mathbf{R}'$) (see symbol description). The latter case can arise if, for example, O' is considered as the point of contact being a rolling cylinder and an inclined plane — the point of contact being the *instantaneous centre*. Moreover, if O' is at rest relative to O or it is chosen to be the centre of mass, we have $\mathbf{R}' \times M\ddot{\mathbf{d}} = \mathbf{0}$ and (6.35) takes the form of (6.11) or (6.19). ☐

6.4   The two-body problem

Consider the motion of an isolated system consisting of two particles. The system is isolated because it is assumed that no external forces act. However, the two particles will influence each other and mutually interact according to Newton's third law. The motion of such a system is referred to as the *two-body problem*.

Our discussion of orbits in Chapter 5 was conducted on the basis that one particle (the gravitational source) was effectively fixed with respect to the orbiting particle. Thence, what amounted to a two-body problem was actually reduced to a one-body problem. This is not unreasonable if one considers the gravitational source to be significantly more massive than the orbiting particle. So, the motion of a pair of particles, neither of them fixed, whose masses are not too dissimilar and therefore influence each other must be analysed using the two-body problem approach.

For our two-particle system, let P_1 be the particle with mass m_1, positioned at $\mathbf{r}_1$, and P_2 be the particle with mass m_2 positioned at $\mathbf{r}_2$, as depicted in Figure 6.5. The centre of mass is positioned at $\mathbf{R}$ somewhere along the line joining P_1 and P_2, where $\overrightarrow{P_2 P_1} = \mathbf{r}$.

The position of the centre of mass for the N-particle case was given by (6.3). Similarly, we write for the two-particle system

$$\mathbf{R} = \frac{m_1 \mathbf{r}_1 + m_2 \mathbf{r}_2}{m_1 + m_2} = \frac{1}{M}(m_1 \mathbf{r}_1 + m_2 \mathbf{r}_2), \tag{6.36}$$

where $M = m_1 + m_2$. The relative position of P_1 with respect to P_2 is given by

$$\mathbf{r} = \mathbf{r}_1 - \mathbf{r}_2. \tag{6.37}$$

Now, the equations of motion for the two-particle system are given by

$$m_1 \ddot{\mathbf{r}}_1 = \mathbf{F}_{12} \tag{6.38}$$

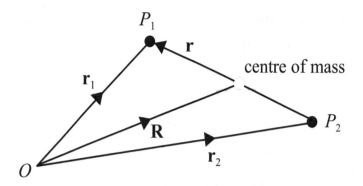

FIGURE 6.5
The two-body problem

and

$$m_2\ddot{\mathbf{r}}_2 = \mathbf{F}_{21}. \tag{6.39}$$

(Recall, we are assuming no external forces are acting on the system.) The force $\mathbf{F}_{12}$ is the force that P_2 exerts on P_1 and *vice versa* for $\mathbf{F}_{21}$.

We are now in a position to reduce our system, consisting of two equations of motion, to one that possesses only one equation of motion. To achieve this, we differentiate (6.37) twice and substitute in (6.38) and (6.39):

$$\ddot{\mathbf{r}} = \ddot{\mathbf{r}}_1 - \ddot{\mathbf{r}}_2 = \frac{\mathbf{F}_{12}}{m_1} - \frac{\mathbf{F}_{21}}{m_2}.$$

On using Newton's third law, we get the *equation of motion of relative position*

$$\mu\ddot{\mathbf{r}} = \mathbf{F}_{12} \equiv \mathbf{F}, \tag{6.40}$$

where $\mathbf{F}_{12} = -\mathbf{F}_{21}$, and the *reduced mass*[7] is defined as

$$\mu = \frac{m_1 m_2}{m_1 + m_2}. \tag{6.41}$$

In going from the pair of equations (6.38) and (6.39) to the single equation (6.40), we have reduced the two-body problem to the equivalent *one-body problem*. In other words, (6.40) is the equation of motion for a single particle of mass μ subject to the force $\mathbf{F}$. So, in solving the one-body problem, we have effectively also solved the two-body problem.

Because there are no external forces acting on the system

$$\ddot{\mathbf{R}} = 0. \tag{6.42}$$

---

[7]The reason μ is called the reduced mass is because it is always less than either m_1 or m_2, and equal to half the mass of either if $m_1 = m_2$.

This implies that the centre of mass moves at a uniform velocity $\dot{\mathbf{R}} = $ constant. Hence, from (6.36), we have the law of conservation of linear momentum:

$$\mathbf{P} = M\dot{\mathbf{R}} = m_1\dot{\mathbf{r}}_1 + m_2\dot{\mathbf{r}}_2. \tag{6.43}$$

Once $\mathbf{R}$ and $\mathbf{r}$ are determined, it is a simple matter to express them in terms of the positions $\mathbf{r}_1$ and $\mathbf{r}_2$. Thus, solving (6.36) and (6.37) simultaneously yields

$$\mathbf{r}_1 = \mathbf{R} + \frac{m_2}{M}\mathbf{r}, \qquad \mathbf{r}_2 = \mathbf{R} - \frac{m_1}{M}\mathbf{r}. \tag{6.44}$$

It is now a straightforward process to write down terms for the kinetic energy (and consequently the total energy) and the angular momentum of the two-body problem, which involve the reduced mass.

For a two-particle system the kinetic energy is defined as

$$
\begin{aligned}
T &= \sum_{i=1}^{2} \frac{1}{2} m_i \dot{\mathbf{r}}_i^2 \\
&= \frac{1}{2} m_1 \dot{\mathbf{r}}_1^2 + \frac{1}{2} m_2 \dot{\mathbf{r}}_2^2 \\
&= \frac{1}{2} m_1 \left(\dot{\mathbf{R}} + \frac{m_2}{M}\dot{\mathbf{r}} \right)^2 + \frac{1}{2} m_2 \left(\dot{\mathbf{R}} - \frac{m_1}{M}\dot{\mathbf{r}} \right)^2 \quad \text{from (6.44)} \\
&= \frac{1}{2} M\dot{\mathbf{R}}^2 + \frac{1}{2}\mu\dot{\mathbf{r}}^2. \tag{6.45}
\end{aligned}
$$

Provided that $\mathbf{F}$ in (6.40) is a central force, the total energy E of the system is

$$E = T + V(\mathbf{r}), \tag{6.46}$$

where $\mathbf{F}(\mathbf{r}) = -\boldsymbol{\nabla}V(\mathbf{r})$.

From (6.9), the angular momentum of the two-particle system is defined as

$$
\begin{aligned}
\mathbf{L} &= \sum_{i=1}^{2} \mathbf{r}_i \times \mathbf{P}_i \\
&= \sum_{i=1}^{2} m_i \mathbf{r}_i \times \dot{\mathbf{r}}_i \\
&= m_1 \mathbf{r}_1 \times \dot{\mathbf{r}}_1 + m_2 \mathbf{r}_2 \times \dot{\mathbf{r}}_2 \\
&= m_1 \left(\mathbf{R} + \frac{m_2}{M}\mathbf{r} \right) \times \left(\dot{\mathbf{R}} + \frac{m_2}{M}\dot{\mathbf{r}} \right) + m_2 \left(\mathbf{R} - \frac{m_1}{M}\mathbf{r} \right) \times \left(\dot{\mathbf{R}} - \frac{m_1}{M}\dot{\mathbf{r}} \right) \\
&= M\mathbf{R} \times \dot{\mathbf{R}} + \mu\mathbf{r} \times \dot{\mathbf{r}} \\
&= \mathbf{R} \times \mathbf{P} + \mu\mathbf{r} \times \dot{\mathbf{r}}. \tag{6.47}
\end{aligned}
$$

Example 6.2 Show that the total energy for the two-body problem is constant.

Solution Differentiating (6.46) with respect to t and using the chain rule yields

$$\frac{dE}{dt} = \frac{dT}{dt} + \frac{\partial V(\mathbf{r})}{\partial \mathbf{r}}\frac{d\mathbf{r}}{dt}$$
$$= M\dot{\mathbf{R}} \cdot \ddot{\mathbf{R}} + \mu\dot{\mathbf{r}} \cdot \ddot{\mathbf{r}} + \boldsymbol{\nabla}V(\mathbf{r}) \cdot \dot{\mathbf{r}}$$
$$= M\dot{\mathbf{R}} \cdot \ddot{\mathbf{R}} + \mathbf{F} \cdot \dot{\mathbf{r}} - \mathbf{F} \cdot \dot{\mathbf{r}}$$
$$= 0.$$

The last line arises due to no external forces acting on the system i.e. $\ddot{\mathbf{R}} = \mathbf{0}$. $\qquad\square$

6.5 Collisions

One particularly important aspect of the two-body problem is the theory of *collisions*. In what follows, we will discuss collisions taking place between two particles in an isolated system. That is, it will be presumed that no external forces will act on the system.

Prior to the collision taking place it will be assumed that the colliding particles will be travelling towards each other at constant velocity having commenced their journey a very great distance apart. During the actual collision itself, the particles may make direct contact with each other (typical in cases where the particles are billiard or pool balls) or be deflected by each other without making direct contact (Rutherford scattering is an example of this type of collision). Following the collision the particles continue with constant velocity. As the particles collide they are under the influence of no other forces except their own mutual forces. The system is isolated so according to Newton's third law the total linear momentum, during the entire collision process, will be conserved. If the total kinetic energy is conserved during the entire collision process, the collision is said to be *elastic*; otherwise, the collision is *inelastic*. It is not, strictly speaking, obvious that elastic collisions between two particles observed in some inertial frame are necessarily elastic when observed in a different inertial frame so that the total kinetic energy before a collision takes place is the same after the collision viewed by two separate inertial observers. To show that this is the case, consider two inertial frames, S and S', where S' is moving at velocity $\mathbf{w}$ relative to S. Denote the velocities of the two particles before the collision by $\mathbf{u}_i$ and the velocities after the collision by $\mathbf{v}_i$, where $i = 1, 2$. So, relative to an observer in S', we have before the collision

$$\mathbf{u}'_i = \mathbf{u}_i - \mathbf{w}$$

and after the collision

$$\mathbf{v}'_i = \mathbf{v}_i - \mathbf{w}.$$

The system is assumed to be isolated, therefore conservation of linear momentum can be employed. Denoting the masses of the two particles by m_i $(i = 1, 2)$ we have

$$\sum_{i=1}^{2} m_i \mathbf{u}_i = \sum_{i=1}^{2} m_i \mathbf{v}_i. \qquad (6.48)$$

Conservation of kinetic energy holds in S, giving

$$\sum_{i=1}^{2} \frac{1}{2} m_i \mathbf{u}_i^2 = \sum_{i=1}^{2} \frac{1}{2} m_i \mathbf{v}_i^2. \qquad (6.49)$$

Let the difference in kinetic energy relative to S' be

$$\begin{aligned}
\Delta T' &= \sum_{i=1}^{2} \frac{1}{2} m_i \mathbf{u}_i'^{\,2} - \sum_{i=1}^{2} \frac{1}{2} m_i \mathbf{v}_i'^{\,2} \\
&= \sum_{i=1}^{2} \frac{1}{2} m_i (\mathbf{u}_i - \mathbf{w}) \cdot (\mathbf{u}_i - \mathbf{w}) - \sum_{i=1}^{2} \frac{1}{2} m_i (\mathbf{v}_i - \mathbf{w}) \cdot (\mathbf{v}_i - \mathbf{w}) \\
&= \sum_{i=1}^{2} \frac{1}{2} m_i \mathbf{u}_i^2 - \sum_{i=1}^{2} \frac{1}{2} m_i \mathbf{v}_i^2 - \sum_{i=1}^{2} m_i \mathbf{u}_i \cdot \mathbf{w} + \sum_{i=1}^{2} m_i \mathbf{v}_i \cdot \mathbf{w} \\
&= \left(\sum_{i=1}^{2} m_i (\mathbf{v}_i - \mathbf{u}_i) \right) \cdot \mathbf{w} \quad \text{from (6.49)} \\
&= 0 \quad \text{from (6.48)}.
\end{aligned}$$

Thus the conservation of kinetic energy also holds in S'.

It should be understood that the above argument says nothing about kinetic energy being conserved throughout the *entire* collision process, only that it is conserved for different inertial observers. In fact, collisions involving billiard balls will release some kinetic energy in the form of heat; nevertheless, these collisions are very close to being elastic. On the other hand, atomic and nuclear particle collisions will have some kinetic energy absorbed during the process.

It appears from our discussions so far that one inertial frame is as good as any other when it comes to analysing an elastic collision process. However, we will see presently that a judicious choice of frame will prove to be more convenient for our analysis.

Before we embark upon this, let us first consider a two-particle collision in an arbitrary inertial frame. The particles are labelled P_1 and P_2 and have masses m_1 and m_2, respectively. The desire is to set up a system of equations involving conservation of linear momentum and conservation of kinetic energy and then solve them for a particular collision problem. The initial (before collision) quantities and the final (after collision) quantities are tabulated in Table 6.1.

TABLE 6.1

Initial and final quantities in a two-particle elastic collision

particle	P_1	P_2
mass	m_1	m_2
initial velocity	$\mathbf{u}_1$	$\mathbf{u}_2$
final velocity	$\mathbf{v}_1$	$\mathbf{v}_2$
initial momentum	$\mathbf{p}_1 = m_1\mathbf{u}_1$	$\mathbf{p}_2 = m_2\mathbf{u}_2$
final momentum	$\mathbf{q}_1 = m_1\mathbf{v}_1$	$\mathbf{q}_2 = m_2\mathbf{v}_2$
initial kinetic energy	$\frac{p_1^2}{2m_1}$	$\frac{p_2^2}{2m_2}$
final kinetic energy	$\frac{q_1^2}{2m_1}$	$\frac{q_2^2}{2m_2}$

From conservation of linear momentum

$$\mathbf{p}_1 + \mathbf{p}_2 = \mathbf{q}_1 + \mathbf{q}_2. \tag{6.50}$$

From conservation of kinetic energy

$$\frac{p_1^2}{m_1} + \frac{p_2^2}{m_2} = \frac{q_1^2}{m_1} + \frac{q_2^2}{m_2}. \tag{6.51}$$

The vector equation (6.50) constitutes three equations. Together with the single equation (6.51), there are a total of four equations. There are four linear momenta each having three components; combined with the two masses, there is a total of 14 quantities. The system is, therefore, *under-determined* — it has fewer equations than unknown variables. It would require more than a given set of initial conditions to determine the final result of the collision. Thus a complete analysis is possible only if 10 of the unknown quantities are given; the remaining four will then be uniquely determined.

6.5.1 Elastic collisions in the laboratory (LAB) frame

The inertial frame in which actual experimental measurements are made is called the *laboratory (LAB) frame*. Although this frame is an appropriate frame for measurements it is not necessarily the most convenient frame. This will become evident in the next section.

In the LAB frame, it is common procedure for one of the two particles to be initially at rest. This is perfectly acceptable as our inertial frame is merely 'moving' at a uniform velocity $\mathbf{u}_2 = 0$, say, if P_2 is the particle at rest. So let us consider P_2 to be at rest, and as such is referred to as the *'target particle'*. The particle moving towards the target particle is referred to as the *'incident particle'* and is travelling at constant initial velocity $\mathbf{u}_1$. The incident particle will be denoted by P_1. The initial and final quantities are then given by Table 6.1. Thence, from (6.50) and (6.51)

$$\mathbf{p}_1 = \mathbf{q}_1 + \mathbf{q}_2 \tag{6.52}$$

and

$$\frac{p_1^2}{m_1} = \frac{q_1^2}{m_1} + \frac{q_2^2}{m_2}. \tag{6.53}$$

A typical LAB frame collision is depicted in Figure 6.6. That all the particles are restricted to a plane is evident from the vector relation (6.52).

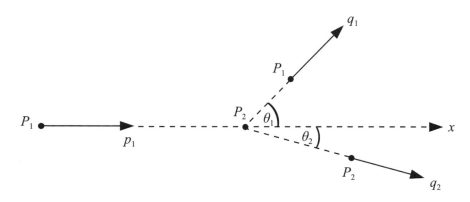

FIGURE 6.6
P_1 collides with P_2 in the LAB frame. After the collision P_1 scatters through θ_1 and P_2 recoils through θ_2

After the collision, P_1 is *scattered* through an angle θ, whilst P_2 *recoils* through an angle θ_2. The components of (6.52) parallel and perpendicular to the x-axis in Figure 6.6 are given, respectively, by

$$p_1 = q_1 \cos \theta_1 + q_2 \cos \theta_2 \tag{6.54}$$

and

$$0 = q_1 \sin \theta_1 - q_2 \sin \theta_2. \tag{6.55}$$

It will also be useful to write (6.53) as

$$q_2^2 = \frac{m_2}{m_1}(p_1^2 - q_1^2). \tag{6.56}$$

Our system has now been reduced to a set of three equations ((6.54), (6.55) and (6.56)) with seven unknown scalar quantities. Thus, for example, given m_1, m_2, p_1 and θ_1, the remaining quantities q_1, q_2 and θ_2 can be determined. Let us now analyse this particular scenario in detail.

Rewriting (6.54) and (6.55) as

$$p_1 - q_1 \cos \theta_1 = q_2 \cos \theta_2$$

and

$$q_1 \sin \theta_1 = q_2 \sin \theta_2,$$

on squaring and adding, θ_2 is eliminated:

$$p_1^2 + q_1^2 - 2p_1q_1\cos\theta_1 = q_2^2.$$

Substituting this into (6.56) yields

$$p_1^2\left(1 - \frac{m_2}{m_1}\right) + q_1^2\left(1 + \frac{m_2}{m_1}\right) - 2p_1q_1\cos\theta_1 = 0.$$

After a little algebra, we get

$$q_1 = \frac{m_1p_1}{m_1 + m_2}\left[\cos\theta_1 \pm \sqrt{\cos^2\theta_1 - \left(1 - \frac{m_2}{m_1}\right)\left(1 + \frac{m_2}{m_1}\right)}\right]. \qquad (6.57)$$

It is necessarily so that q_1 is real and positive since $q_1 = |q_1|$. This implies that the discriminant in (6.57) is non-negative. That is

$$\cos^2\theta_1 - \left(1 - \frac{m_2}{m_1}\right)\left(1 + \frac{m_2}{m_1}\right) \geq 0$$

or

$$\cos^2\theta_1 \geq \left(1 - \frac{m_2}{m_1}\right)\left(1 + \frac{m_2}{m_1}\right). \qquad (6.58)$$

If $m_1 > m_2$ then from (6.58), we must have

$$\cos^2\theta_1 \geq \left(1 - \frac{m_2}{m_1}\right)\left(1 + \frac{m_2}{m_1}\right) \qquad (6.59)$$

or

$$\cos^2\theta_1 \leq -\left(1 - \frac{m_2}{m_1}\right)\left(1 + \frac{m_2}{m_1}\right). \qquad (6.60)$$

The last inequality (6.60) is impossible as it suggests that $q_1 < 0$. As m_2/m_1 lies within the range $0 < m_2/m_1 < 1$, we have from (6.59) $0 < \theta_1 < \pi/2$. Hence, there is a maximum scattering angle for P_1. If $m_1 \gg m_2$ then the scattering angle will be small, as would be expected.

If $m_1 < m_2$ then all scattering angles are possible. That is $0 \leq \theta_1 \leq \pi$. In this case the positive square root must be chosen. If $\theta_1 = 0$ then from (6.57) $q_1 = p_1$, which implies that a collision is not possible. If $\theta_1 = \pi$ then from (6.57) $q_1 = p_1(m_2 - m_1)/(m_2 + m_1)$ and the collision is head-on.

If $m_1 = m_2$, we have from (6.57)

$$q_1 = p_1\cos\theta_1. \qquad (6.61)$$

From (6.54), (6.55) and (6.56) we have

$$q_2 = p_1\sin\theta_1 \qquad (6.62)$$

and

$$\theta_1 + \theta_2 = \frac{\pi}{2}. \qquad (6.63)$$

The range of θ_1 is now $0 \leq \theta_1 < \pi/2$. If $\theta_1 = 0$ then $q_1 = p_1$ and a collision does not take place. As $\theta_1 \to \pi/2$, $q_1 \to 0$ and the collision is approximately head-on. In this case nearly all of the momentum of P_1 is transferred to P_2.

6.5.2 Elastic collisions in the centre of mass (CM) frame

Although the LAB frame is the frame in which experimental measurements are made, it is not necessarily the most convenient frame. Indeed, a somewhat nicer frame to discuss particle collisions is the *centre of mass (CM) frame*.[8] In this frame, the centre of mass of the two-particle system is at rest.

That the CM frame is a much more tractable frame in which to perform calculations will shortly become evident when we consider linear momentum and kinetic energy conservation during the elastic collision process.

It has already been shown that for an N-particle system the total linear momentum relative to the centre of mass is zero. Hence, appealing to (6.15) for a two-particle system, the total momentum before collision is

$$\mathbf{P}^c = \sum_{i=1}^{2} m_i \dot{\mathbf{r}}_i^c$$
$$= m_1 \dot{\mathbf{r}}_1^c + m_2 \dot{\mathbf{r}}_2^c$$
$$= \mathbf{p}_1^c + \mathbf{p}_2^c = 0. \tag{6.64}$$

This implies

$$\mathbf{p}_1^c = -\mathbf{p}_2^c \equiv \mathbf{p}^c. \tag{6.65}$$

But linear momentum is conserved during the entire collision process:

$$\mathbf{p}_1^c + \mathbf{p}_2^c = \mathbf{q}_1^c + \mathbf{q}_2^c, \tag{6.66}$$

where $\mathbf{q}_1^c$ and $\mathbf{q}_2^c$ are the linear momenta after the collision in the CM frame. Hence,

$$\mathbf{q}_1^c + \mathbf{q}_2^c = \mathbf{0}. \tag{6.67}$$

This implies

$$\mathbf{q}_1^c = -\mathbf{q}_2^c \equiv \mathbf{q}^c. \tag{6.68}$$

From (6.65), we have

$$(p_1^c)^2 = (p_2^c)^2 \equiv (p^c)^2. \tag{6.69}$$

So, in the CM frame both particles are initially approaching each other with the same magnitude of momentum. From (6.68), we have

$$(q_1^c)^2 = (q_2^c)^2 \equiv (q^c)^2. \tag{6.70}$$

After the collision both particles are receding from each other with the same magnitude of momentum. This is depicted in Figure 6.7. The angle θ^c is the scattering angle.

---

[8]This is also referred to as the *centre of momentum frame* or *zero momentum frame*.

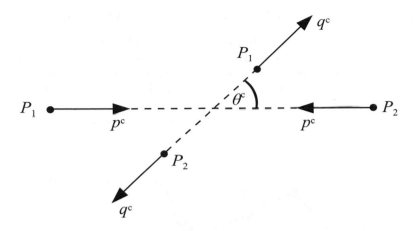

FIGURE 6.7
Two-particle collision in the CM frame

The total kinetic energy before collision is, using (6.22),

$$T^c = \sum_{i=1}^{2} \frac{1}{2} m_i (\dot{\mathbf{r}}_i^c)^2$$
$$= \frac{1}{2} m_1 (\dot{\mathbf{r}}_1^c)^2 + \frac{1}{2} m_2 (\dot{\mathbf{r}}_2^c)^2$$
$$= \frac{(p_1^c)^2}{2m_1} + \frac{(p_2^c)^2}{2m_2}.$$

Because the collision process is elastic, conservation of kinetic energy can be employed. Hence,

$$T^c = \frac{(p_1^c)^2}{2m^1} + \frac{(p_2^c)^2}{2m^2} = \frac{(q_1^c)^2}{2m_1} + \frac{(q_2^c)^2}{2m_2},$$

or, with the aid of (6.69) and (6.70),

$$T^c = \frac{(p^c)^2}{2} \left(\frac{m_1 + m_2}{m_1 m_2} \right) = \frac{(q^c)^2}{2} \left(\frac{m_1 + m_2}{m_1 m_2} \right)$$
$$= \frac{(p^c)^2}{2\mu} = \frac{(q^c)^2}{2\mu}, \tag{6.71}$$

where μ is the reduced mass of the system (see (6.41)). It follows, therefore, that

$$p^c = q^c. \tag{6.72}$$

So, the magnitudes of the momenta for each particle P_1 and P_2 before and after colliding are the same.

6.5.3   Transformations between CM and LAB frames

In spite of the fact that a great deal of economy in calculation can be gained by using the CM frame, at some point we nevertheless must revert back to the LAB frame due to the practicalities of experimental observation.

First, let us consider how the momenta transform between the CM and LAB frames. Figure 6.8 shows how particles P_1 and P_2 are positional relative to the centre of mass. The position of the centre of mass $\mathbf{R}$ is given by (6.36).

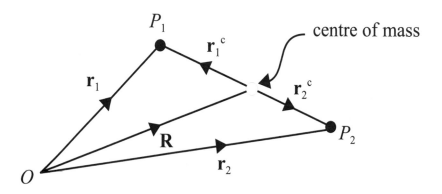

FIGURE 6.8
P_1 and P_2 relative to the centre of mass

Differentiating this yields

$$\dot{\mathbf{R}} = \frac{m_1\dot{\mathbf{r}}_1 + m_2\dot{\mathbf{r}}_2}{M}$$
$$= \frac{\mathbf{p}_1 + \mathbf{p}_2}{M},$$

where $\mathbf{p}_1$ and $\mathbf{p}_2$ are the before collision momenta in the LAB frame (see Table 6.1). But $\mathbf{p}_2 = \mathbf{0}$ in the LAB frame, implying that

$$\dot{\mathbf{R}} = \frac{\mathbf{p}_1}{M}, \tag{6.73}$$

where $M = m_1 + m_2$. From Figure 6.8 we can write

$$\mathbf{r}_1 = \mathbf{R} + \mathbf{r}_1^c \tag{6.74}$$

and

$$\mathbf{r}_2 = \mathbf{R} + \mathbf{r}_2^c. \tag{6.75}$$

Differentiating (6.74) and multiplying through by m_1 yields

$$\mathbf{p}_1 = m_1\dot{\mathbf{R}} + \mathbf{p}_1^c = m_1\dot{\mathbf{R}} + \mathbf{p}^c \tag{6.76}$$

on employing (6.65). Differentiating (6.75) and multiplying through by m_2 yields

$$\mathbf{p}_2 = m_2\dot{\mathbf{R}} + \mathbf{p}_2^c = m_2\dot{\mathbf{R}} - \mathbf{p}^c$$

on employing (6.65). However, $\mathbf{p}_2 = \mathbf{0}$ in the LAB frame. Thus

$$\mathbf{p}^c = m_2\dot{\mathbf{R}}. \tag{6.77}$$

Substituting (6.77) into (6.76) gives

$$\mathbf{p}_1 = \frac{M}{m_2}\mathbf{p}^c. \tag{6.78}$$

After the collision takes place the velocity of the centre of mass is given by

$$\dot{\mathbf{R}} = \frac{m_1\mathbf{v_1} + m_2\mathbf{v_2}}{M} = \frac{\mathbf{q_1} + \mathbf{q_2}}{M},$$

where

$$\mathbf{v}_1 = \dot{\mathbf{R}} + \mathbf{v}_1^c \tag{6.79}$$

and

$$\mathbf{v}_2 = \dot{\mathbf{R}} + \mathbf{v}_2^c. \tag{6.80}$$

Multiplying (6.79) by m_1 yields

$$\mathbf{q}_1 = m_1\dot{\mathbf{R}} + \mathbf{q}_1^c = m_1\dot{\mathbf{R}} + \mathbf{q}^c \tag{6.81}$$

on employing (6.68). Multiplying (6.80) by m_2 yields

$$\mathbf{q}_2 = m_2\dot{\mathbf{R}} + \mathbf{q}_2^c = m_2\dot{\mathbf{R}} - \mathbf{q}^c \tag{6.82}$$

on employing (6.68). Substituting (6.73) into (6.81) and (6.82) yields

$$\mathbf{q}_1 = \frac{m_1}{m_2}\mathbf{p}^c + \mathbf{q}^c \tag{6.83}$$

and

$$\mathbf{q}_2 = \mathbf{p}^c - \mathbf{q}^c. \tag{6.84}$$

Together with (6.78), equations (6.83) and (6.84) are the momenta transformations between the CM and LAB frames.

A useful way of extracting further information is to construct a vector triangle involving the momenta in (6.78), (6.83) and (6.84). This is depicted in Figure 6.9.

From (6.72) $|\mathbf{q}^c| = |\mathbf{p}^c|$; therefore, triangle CEB is isosceles. Hence, the recoil angle θ_2 is related to the CM frame scattering angle θ^c by

$$\theta_2 = \frac{\pi - \theta^c}{2}. \tag{6.85}$$

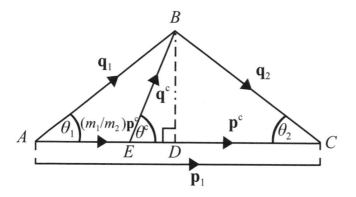

FIGURE 6.9
Representation of momenta transformations between the CM and LAB frames

Now, it is easily seen using simple trigonometry that

$$\tan \theta_1 = \frac{BD}{(m_1/m_2)|\mathbf{p}^c| + ED}.$$

Also,

$$\sin \theta^c = \frac{BD}{|\mathbf{q}^c|}, \quad \cos \theta^c = \frac{ED}{|\mathbf{q}^c|}.$$

Hence, the scattering angle θ_1 in the LAB frame is related to the scattering angle θ^c in the CM frame by

$$\tan \theta_1 = \frac{\sin \theta^c}{(m_1/m_2) + \cos \theta^c}. \tag{6.86}$$

We can use (6.86) to establish a few interesting facts about the particle masses and the scattering angles in the CM and LAB frames. It was established in Section 6.5.1 that there is a maximum scattering angle for P_1 in the LAB frame. We will call this maximum angle $\theta_1 = \theta_{\max}$. This maximum scattering angle can only be found by differentiating (6.86) with respect to θ and equating the derivative to zero. Hence,

$$\frac{d}{d\theta}(\tan \theta_1) = \frac{(m_1/m_2)\cos \theta^c + 1}{((m_1/m_2) + \cos \theta^c)^2} = 0$$

implying that $\theta_1 = \theta_{\max}$ when $\cos \theta^c = -m_2/m_1$. This is possible only if $m_1 > m_2$. We can write (6.86) as

$$\tan \theta_{\max} = \left(\frac{m_1}{m_2} - 1\right)^{-\frac{1}{2}} \left(\frac{m_1}{m_2} + 1\right)^{-\frac{1}{2}}.$$

Using the identity $\cos^2\theta(1+\tan^2\theta) = 1$, we find that

$$\cos\theta_{\max} = \left(1 - \frac{m_2}{m_1}\right)^{\frac{1}{2}}\left(1 + \frac{m_2}{m_1}\right)^{\frac{1}{2}} \tag{6.87}$$

and, therefore,

$$\sin\theta_{\max} = \frac{m_2}{m_1}.$$

Notice that (6.87) corresponds to (6.59) when equality is observed.
    If $m_1 = m_2$ then from (6.86),

$$\tan\theta_1 = \frac{\sin\theta^c}{1 + \cos\theta^c} = \frac{2\sin(\theta^c/2)\cos(\theta^c/2)}{2\cos^2(\theta^c/2)} = \tan\left(\frac{\theta^c}{2}\right)$$

so

$$\theta_1 = \frac{\theta^c}{2},$$

where the maximum value of θ_1 occurs at a scattering angle of $\pi/2$. This is in agreement with the result previously attained in Section 6.5.1.

Example 6.3 For a two-particle system in the LAB frame, find the sum of the scattering angle and the recoil angle in terms of the particle masses m_1 and m_2 and the scattering angle in the CM frame.

Solution The sum of the two angles in question is θ_1 and θ_2 (see Figure 6.6). Thus we can write

$$\tan(\theta_1 + \theta_2) = \frac{\tan\theta_1 + \tan\theta_2}{1 - \tan\theta_1\tan\theta_2}.$$

Substituting (6.85) and (6.86) into the equation yields

$$\tan(\theta_1 + \theta_2) = \frac{\sin\theta^c + \left(\left(\frac{m_1}{m_2}\right) + \cos\theta^c\right)\lim_{\alpha\to\pi/2}\tan\frac{1}{2}(\alpha - \theta^c)}{\frac{m_1}{m_2} + \cos\theta^c - \sin\theta^c\lim_{\alpha\to\pi/2}\tan\frac{1}{2}(\alpha - \theta^c)}.$$

Now,

$$\lim_{\alpha\to\pi/2}\tan\frac{1}{2}(\alpha - \theta^c) = \lim_{\alpha\to\pi/2}\frac{\tan(\alpha/2) - \tan(\theta^c/2)}{1 + \tan(\alpha/2)\tan(\theta^c/2)}$$

$$= \cot\left(\frac{\theta^c}{2}\right)$$

using L'Hospital's rule (see Exercise 8 in Chapter 3 for a definition). Therefore,

$$\tan(\theta_1 + \theta_2) = \frac{2\sin(\theta^c/2)\cos(\theta^c/2) + \left(\frac{m_1}{m_2} + 1 - 2\sin^2(\theta^c/2)\right)\cot(\theta^c/2)}{\frac{m_1}{m_2} + 1 - 2\sin^2(\theta^c/2) - 2\sin(\theta^c/2)\cos(\theta^c/2)\cot(\theta^c/2)}.$$

After simplifying and taking the inverse tan the result follows:

$$\theta_1 + \theta_2 = \tan^{-1}\left[\left(\frac{m_1}{m_2}+1\right)\left(\frac{m_1}{m_2}-1\right)^{-1}\cot\left(\frac{\theta^c}{2}\right)\right]. \qquad \square$$

Example 6.4 For the two-particle collision depicted in Figure 6.6, find the ratio $T_{\mathbf{q}2}/T_{\mathbf{p}1}$, where $T_{\mathbf{q}2}$ is the kinetic energy of the target particle after the collision, and $T_{\mathbf{p}1}$ is the kinetic energy of the incident particle before the collision. Give your answer in terms of the CM frame scattering angle. When is this energy transfer a maximum? What percentage of maximum energy is transferred if the mass of the incident particle m_1 is related to the mass of the target particle m_2 by $m_1 = 3m_2$?

Solution From Table 6.1 the kinetic energy $T_{\mathbf{q}2}$ is given by

$$T_{\mathbf{q}2} = \frac{q_2^2}{2m_2}.$$

By appealing directly to (6.84) or Figure 6.9, we have

$$q_2^2 = 2p^{c\,2}(1 - \cos\theta^c).$$

Thus

$$T_{\mathbf{q}2} = \frac{p^{c\,2}}{m_2}(1 - \cos\theta^c).$$

The total kinetic energy before the collision is supplied only by the incident particle. Thus, from Table 6.1,

$$T_{\mathbf{p}1} = \frac{p_1^2}{2m_1} = \frac{1}{2m_1}\frac{M^2}{m_2^2}p^{c\,2}$$

on employing (6.78). Hence,

$$\frac{T_{\mathbf{q}2}}{T_{\mathbf{p}1}} = \frac{2\mu}{M}(1 - \cos\theta^c),$$

where μ is the reduced mass and $M = m_1 + m_2$.

The energy transfer is maximum when $\theta^c = \pi$ (this corresponds to a head-on collision). Thence,

$$\frac{T_{\mathbf{q}2}}{T_{\mathbf{p}1}} = \frac{4\mu}{M}.$$

For the stipulated mass ratio $(m_1/m_2 = 3)$

$$\frac{T_{\mathbf{q}2}}{T_{\mathbf{p}1}} = \frac{3}{4}.$$

Therefore, 75 percent of the maximum energy is transferred during the collision. $\square$

Example 6.5 A ball of mass m_1 is suspended fractionally above a ball of mass m_2 so that a line joining their centres of mass is precisely vertical (see Figure 6.10). Assume that $m_1 < m_2$. The balls

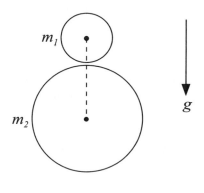

FIGURE 6.10
Collision between two balls falling vertically

are released simultaneously so that just before they collide with the ground their speed is u. Assuming that all subsequent collisions are elastic, use conservation of energy and momentum to determine the speed v of the smaller ball after both balls collide.

After both balls collide, the smaller ball rises to a height of 9 m. If $m_1 = 0.1$ kg and $m_2 = 0.7$ kg, find the speed of the balls just prior to the larger ball's collision with the ground. Take g to be 10 ms^{-2}.

Solution Because the collisions are elastic the bigger ball will be travelling at speed u in the opposite direction to the smaller moments before they collide. So using conservation of energy and momentum, we have

$$m_1 u^2 + m_2 u^2 = m_1 v^2 + m_2 w^2$$

and

$$-m_1 u + m_2 u = m_1 v + m_2 w,$$

where we take the upwards direction as positive. Our goal is to eliminate w and write v in terms of u. Now, writing the energy equation as

$$m_1 (u - v)(u + v) = m_2 (w - u)(w + u)$$

and the momentum equation as

$$-m_1 (u + v) = m_2 (w - u)$$

and dividing, yields

$$-(u - v) = (w + u) \Rightarrow w = v - 2u.$$

So, on eliminating w from the momentum equation, we have the desired result:

$$v = u \frac{(3m_2 - m_1)}{m_2 + m_1}.$$

Now, after the balls collide, the smaller ball will be travelling upwards at speed $v_{\text{initial}} = v$. It will reach its maximum height when its final velocity v_{final} is zero. Hence with the aid of the kinematic relation

$$v_{\text{final}}^2 - v_{\text{initial}}^2 = -2gs,$$

where s is vertical displacement, we have

$$u^2 = 2gs \left(\frac{m_2 + m_1}{3m_2 - m_1} \right)^2.$$

Substituting in the given values yields

$$u = \sqrt{2(10)(9)} \left(\frac{0.1 + 0.7}{3(0.7) - 0.1} \right) = 5.37.$$

Therefore, the speed of the balls prior to the larger ball colliding with the ground is 5.37 ms^{-1}. $\square$

6.6   Inelastic collisions

As was mentioned earlier, the collision process generally involves some release or absorption of energy. However, during the collision process kinetic energy remains conserved provided that a balancing term is incorporated in the energy conservation equation.

The release or absorption of energy is probably most manifest in elementary particle collisions. This invariably gives rise to the production of 'new' particles from the colliding 'old' particles. For example, in sufficiently energetic proton-proton collisions a neutral pion (pi-meson) may be produced.

In light of all this, the conservation of energy equations in LAB and CM frames must be recast as follows. In the LAB frame

$$\frac{p_1^2}{m_1} = \frac{q_3^2}{m_3} + \frac{q_4^2}{m_4} + 2Q \tag{6.88}$$

and in the CM frame

$$\frac{(p_1^c)^2}{m_1} + \frac{(p_2^c)^2}{m_2} = \frac{(q_3^c)^2}{m_3} + \frac{(q_4^c)^2}{m_4} + 2Q, \tag{6.89}$$

where Q is the *kinetic energy balancing term*[9], q_3 and q_4 are the final momenta and m_3 and m_4 are the final masses.

Conservation of linear momentum in the CM frame implies that

$$(p_1^c)^2 = (p_2^c)^2 \equiv (p^c)^2$$

and

$$(q_3^c)^2 = (q_4^c)^2 \equiv (q^c)^2$$

(refer to (6.69) and (6.70)), so (6.89) can be written as

$$\frac{(p^c)^2}{\mu} = \frac{(q^c)^2}{\mu'} + 2Q, \tag{6.90}$$

where

$$\mu = \left(\frac{m_1 m_2}{m_1 + m_2}\right) \text{ and } \mu' = \left(\frac{m_3 m_4}{m_3 + m_4}\right)$$

are the reduced masses of the 'old' particle pair (before collision) and the 'new' particle pair (after collision).

If $Q > 0$ then energy is released during a collision, and if $Q < 0$ energy is absorbed. The generic term *inelastic* is used to describe collision processes for $Q \neq 0$. Of course, if $Q = 0$ the collision would be elastic with $m_1 = m_3$ and $m_2 = m_4$.

Example 6.6 A pair of particles P_1 and P_2 with masses m_1 and m_2 collide inelastically in the LAB frame. After the collision, two particles P_3 and P_4 with masses m_3 and m_4 are produced. If the momenta of P_1, P_2, P_3 and P_4 are $p_1, 0, q_3$ and q_4, respectively, obtain a relation for Q in terms of α, where α is the angle displayed in Figure 6.11.

Solution From conservation of linear momentum

$$p_1 - q_3 \cos \alpha = q_4 \cos \beta$$

and

$$q_3 \sin \alpha = q_4 \sin \beta.$$

On squaring and adding, β can be eliminated:

$$p_1^2 + q_3^2 - 2p_1 q_3 \cos \alpha = q_4^2.$$

---

[9]Notice that Q is frame independent. This is because the relative velocities of the particles before and after a collision are frame independent.

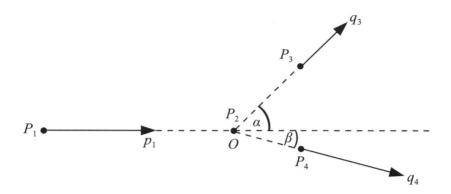

FIGURE 6.11
Production of a pair of particles arising from a collision in the LAB frame

Combining this with the conservation of energy equation (6.88) yields, after a little algebra,

$$2Q = p_1^2 \left(\frac{1}{m_1} - \frac{1}{m_4} \right) - q_3^2 \left(\frac{1}{m_3} + \frac{1}{m_4} \right) + \frac{2 p_1 q_3 \cos \alpha}{m_4}. \qquad \square$$

Example 6.7 A pair of particles of equal mass collide inelastically. One particle is at rest while the other particle travels with linear momentum $\mathbf{p}_1$ in the LAB frame. Find the initial momenta in the CM frame and the energy released during the collision if the final momentum is zero as measured in the CM frame. What is the final momentum in the LAB frame if the particles coalesce into a a single particle of mass $2m$?

Solution As the particles are of equal mass, the initial momentum in the CM frame is given by (6.78):

$$\mathbf{p}^c = \frac{1}{2}\mathbf{p}_1.$$

On using (6.65), we arrive at

$$\mathbf{p}_1^c = \frac{1}{2}\mathbf{p}_1$$

and

$$\mathbf{p}_2^c = -\frac{1}{2}\mathbf{p}_1.$$

In the CM frame $\mathbf{q}^c = \mathbf{0}$. Employing (6.90) yields

$$Q = \frac{(\mathbf{p}^c)^2}{2\mu} = \frac{(\mathbf{p}_1^c)^2}{m},$$

where $\mu = m/2$. This is the total energy released during the collision.

When the particles coalesce, they will form a single particle of mass $2m$ travelling with momentum $\mathbf{q}_{34}$, say, in the LAB frame. So, the conservation of energy equation (6.88) can be written as

$$\frac{\mathbf{q}_{34}^2}{2m} = \mathbf{p}_1^2 - 2Q$$

or

$$\mathbf{q}_{34}^2 = 2\mathbf{p}_1^2 - \mathbf{p}_1^2$$

implying that

$$\mathbf{q}_{34} = \mathbf{p}_1. \qquad \square$$

In the case when $\mu = \mu'$ (6.90) takes the simple form

$$(p^c)^2 - (q^c)^2 = 2\mu Q.$$

Thus the difference in the squares of the magnitudes of the initial and final momenta in the CM frame is directly proportional to the energy absorbed during the collision. The constant of proportionality is written either in terms of the masses before or after the collision.

6.6.1 The coefficient of restitution

Another way of discussing the release of energy during inelastic collisions is with the empirically determined quantity e called the *coefficient of restitution*. Discovered by Newton, through experiments with two non-rotating bodies colliding head-on, the coefficient of restitution is the magnitude of the relative speed after collision divided by the relative speed before collision along the line of collision. It can be written as

$$v_2 - v_1 = e(u_1 - u_2). \qquad (6.91)$$

(In actuality, e is not really a constant as it depends on, amongst other things, the initial velocities of the particles and the medium in which the collisions occur.)

Notice that e is frame independent. It has an upper and lower limit: $0 \leq e \leq 1$. For $e = 0$, the collision is completely inelastic (or *plastic*) — the particles coalesce after the collision and are at rest in the CM frame. For $e = 1$, the collision is completely elastic — no energy is released and $Q = 0$.

If the particle masses are unaffected by the collision process, a relation between Q and e can be determined.

Consider the inelastic collision between two particles in the LAB frame. The initial velocity — and therefore the initial momentum of the second particle, say, is zero. Conservation of linear momentum gives

$$p_1 = q_1 + q_2 \qquad (6.92)$$

and (6.91) becomes. when written as a relation in terms of particle momenta,

$$\frac{q_2}{m_2} - \frac{q_1}{m_1} = \frac{ep_1}{m_1}.$$ (6.93)

Combining (6.92) and (6.93) yields

$$q_1 = p_1 \left(\frac{m_1 - em_2}{m_1 + m_2} \right), \quad q_2 = p_1 \left(\frac{m_2(1 + e)}{m_1 + m_2} \right).$$ (6.94)

From (6.88), the energy released by the collision satisfies

$$2Q = \frac{p_1^2}{m_1} - \left(\frac{q_1^2}{m_1} + \frac{q_2^2}{m_2} \right),$$ (6.95)

which amounts to the difference between initial and final kinetic energy. Combining (6.94) and (6.95) yields

$$2Q = \frac{\mu(1 - e^2)p_1^2}{m_1^2},$$

where μ is the reduced mass.

We can also deduce from this that e must be less than unity for energy to be released during the collision.

Example 6.8 Two particles P_1 and P_2 with equal mass collide inelastically in the LAB frame. P_2 is at rest. As they collide, the straight line AB joining the particles makes an angle $\pi/6$ with the velocity of P_1. If the scattering angle is α, determine a value for the coefficient of restitution. Consult Figure 6.12.

Solution Using conservation of linear momentum parallel and perpendicular to AB yields

$$\rightarrow \quad u_1 = v_1 \cos \alpha + v_2 \cos(\pi/6)$$
$$\uparrow \quad 0 = -v_1 \sin \alpha + v_2 \sin(\pi/6).$$

From the equation of restitution (6.91)

$$v_2 \cos(\pi/6) - v_1 \cos \alpha = eu_1.$$

After evaluating the trigonometric function, the three equations can be written as

$$u_1 = v_1 \cos \alpha + \frac{\sqrt{3}}{2} v_2$$

$$0 = -v_1 \sin \alpha + \frac{1}{2} v_2$$

$$eu_1 = -v_1 \cos \alpha + \frac{\sqrt{3}}{2} v_2.$$

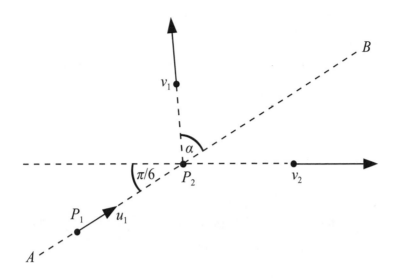

FIGURE 6.12
P_1 and P_2 collide obliquely

Replacing v_2 in the first and third equations by v_2 in the second equation and combining the results gives

$$e(\sqrt{3}v_1 \sin \alpha + v_1 \cos \alpha) = \sqrt{3}v_1 \sin \alpha - v_1 \cos \alpha.$$

On cancelling v_1, we finally arrive at

$$e = \frac{\sqrt{3}\tan \alpha - 1}{\sqrt{3}\tan \alpha + 1}. \qquad \square$$

6.7 Variable mass problems

There are many examples whereby a body moving in an inertial frame may have its mass vary with respect to time. That is, the body may lose or gain mass over time. Two examples that very clearly illustrate the variable mass problem are rocket motion and falling raindrops. They will both be considered separately.

6.7.1 Rocket motion

The motion of a rocket through air or space *is* a multi-particle system. In fact, it can be treated as a two-particle system if the rocket and the mass of exhaust gas expelled by the rocket are considered as separate bodies. If the rocket motion is influenced by external forces then Newton's second law in the form of (6.6) applies.

We first derive the rocket equation and then discuss some possible solutions.

Consider a rocket at time t of mass $m(t)$ travelling in a straight line with velocity $v(t)$ (see Figure 6.13). At time $t + \delta t$ the rocket expels exhaust gas at

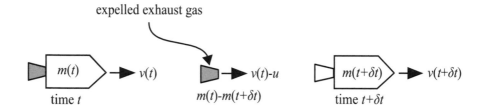

FIGURE 6.13
Schematic representation of a rocket expelling exhaust gas

speed u relative to the rocket. The mass of the rocket, at this time, is reduced to $m(t + \delta t)$. The mass of the expelled exhaust gas is $m(t) - m(t + \delta t)$ and travels at velocity $v(t) - u$ relative to a fixed inertial frame.

The momentum, at time t, of the rocket before exhaust gas is expelled is

$$p(t) = m(t)v(t).$$

The momentum, at time $t + \delta t$, of the expelled exhaust gas, $p_G(t + \delta t)$, plus the rocket $p_R(t + \delta t)$ is

$$p(t + \delta t) = p_G(t + \delta t) + p_R(t + \delta t),$$

where

$$p_G(t + \delta t) = (m(t) - m(t + \delta t))(v(t) - u) \tag{6.96}$$

and

$$p_R(t + \delta t) = (m(t + \delta t)v(t + \delta t). \tag{6.97}$$

Performing a Taylor expansion on the right of (6.96) yields

$$p_G(t + \delta t) \approx \left(-\delta t \frac{dm(t)}{dt} \right) \left(v(t + \delta t) - \delta t \frac{dv(t)}{dt} - u \right)$$

$$= -\delta t \frac{dm(t)}{dt}(v(t + \delta t) - u) + (\delta t)^2 \frac{dm(t)}{dt} \frac{dv(t)}{dt}.$$

Performing a Taylor expansion on the right of (6.97) yields

$$p_R(t + \delta t) \approx \left(m(t) + \delta t \frac{dm(t)}{dt} \right) \left(v(t) + \delta t \frac{dv(t)}{dt} \right)$$

$$m(t)v(t) + \delta t \left(\frac{dm(t)}{dt} + \frac{dv(t)}{dt} \right) + (\delta t)^2 \frac{dm(t)}{dt} \frac{dv(t)}{dt}.$$

From the definition of the derivative of the function $p(t)$:

$$p'(t) = \lim_{\delta t \to 0} \frac{p(t + \delta t) - p(t)}{\delta t},$$

we have

$$\dot{p}(t) = \lim_{\delta t \to 0} \frac{p_G(t + \delta t) + p_R(t + \delta t) - p(t)}{\delta t}$$

$$= \lim_{\delta t \to 0} \frac{1}{\delta t} \left[-\delta t \frac{dm(t)}{dt} (v(t + \delta t) - u) + \delta t \left(\frac{dm(t)}{dt} + \frac{dv(t)}{dt} \right) \right].$$

Notice that the terms of order $(\delta t)^2$ have been ignored. On taking the limit, we finally arrive at

$$\dot{p}(t) = m(t) \frac{dv(t)}{dt} + u \frac{dm(t)}{dt} = F, \tag{6.98}$$

where F corresponds to any external force that may arise, such as gravity or air resistance. Equation (6.98) is called the *rocket equation*.

Example 6.9 Solve the rocket equation when there is no external force: $F = 0$.

Solution In this case, (6.98) becomes

$$m(t) \frac{dv(t)}{dt} = -u \frac{dm(t)}{dt}.$$

The term $-dm(t)/dt$ is the rate at which the rocket is expelling mass. Because this equation is of the form *mass* × *acceleration* = *force*, the right-hand side is referred to as *thrust*. Since $dm(t)/dt < 0$, the thrust is a positive force.

Assume initial conditions $v(0) = v_0, m(0) = m_0$. We can write, on integrating,

$$\int dv(t) = -u \int d(\ln(m(t))),$$

giving

$$v(t) = v_0 + u \ln \left(\frac{m_0}{m(t)} \right), \tag{6.99}$$

on applying the initial conditions. If the rocket fuel is all used by

$t = t_1$, say, then $v(t_1) = v_1$ and $m(t_1) = m_1$, so the increase in speed
is

$$\Delta v = v_1 - v_0 = u \ln \left(\frac{m_0}{m_1} \right).$$

This equation is sometimes referred to as *Tsiolkovsky's rocket equation*.[10] It places a restriction on the rocket's maximum speed. (For the exaggerated ratio $m_0/m_1 = 10^5$, $\Delta v = 11.5u$, i.e. the overall increase in speed is no greater than 11.5 times the speed of the expelled exhaust gas!) ☐

Example 6.10 Find the distance travelled by the rocket in Example 6.9 if the rate at which the rocket burns fuel is

$$\frac{dm(t)}{dt} = -k, \quad k > 0.$$

Solution Assuming that $m(0) = m_0$, we find on integrating

$$m(t) = m_0 - kt.$$

Substituting this into (6.99) gives

$$v(t) = v_0 - u \ln \left(\frac{m_0 - kt}{m_0} \right),$$

where $t < m_0/k$; otherwise the rocket would eventually vanish! In other words, if $t = m_0/k$ the rocket would have to be composed entirely of fuel.

Assuming that the rocket is travelling in the x-direction and that $x(0) = 0$, we find on integrating

$$x(t) = (v_0 + u)t + \frac{u}{k}(m_0 - kt) \ln \left(\frac{m_0 - kt}{m_0} \right),$$

which is the distance travelled by the rocket. ☐

For the last part of this section, we will consider incorporating an external force in the form of a uniform gravitational field. In this case, (6.98) becomes

$$m(t)\frac{dv(t)}{dt} = -u\frac{dm(t)}{dt} - m(t)g.$$

Note that the terms on the right of this equation correspond to *upwards* thrust

---

[10]It was named after the Russian scientist Konstantin Tsiolkovsky (1857-1935) who published it in 1903. However, allegedly, the equation had already been derived by the British mathematician William Moore in 1813.

and *downwards* weight. Assume that these forces act along the positive and negative z-direction, respectively. On dividing by $m(t)$ and integrating

$$\int dv(t) = -u \int d(\ln(m(t))) - g \int dt.$$

Assuming initial conditions $v(0) = v_0$, $m(0) = m_0$, we have

$$v(t) = v_0 + u \ln \left(\frac{m_0}{m(t)} \right) - gt. \tag{6.100}$$

If the rocket fuel is all used by $t = t_1$, then $v(t_1) = v_1$ and $m(t_1) = m_1$, so the increase in speed is

$$\Delta v = v_1 - v_0 = u \ln \left(\frac{m_0}{m_1} \right) - gt_1.$$

If the rocket expels exhaust gas too slowly, that is if

$$t_1 \geq [v_0 + u \ln((m_0 - \alpha t)/m_0)/g]$$

then

$$\Delta v \leq 0,$$

which means that the rocket will not have the required thrust to leave the launch pad. This can be remedied provided $-udm(t)/dt > m(t)g$.

6.7.2 A falling raindrop

The rocket equation is not only useful for analysing rocket motion, it can be used in most situations where a body's mass varies over time.

In the previous section, we were concerned with a body losing mass; here, we consider the opposite case — that of a body increasing in mass.

Consider a raindrop of mass $m(t)$ falling through the atmosphere at speed $v(t)$. It now falls through a stationary cloud. As it does so, the raindrop increases in mass as it collects moisture from its surroundings. The moisture in the cloud is stationary relative to the cloud, which means that $v(t) - u = 0$. The rocket equation can then be written as

$$m(t)\frac{dv(t)}{dt} + v(t)\frac{dm(t)}{dt} = m(t)g$$

or

$$\frac{dm(t)v(t)}{dt} = m(t)g. \tag{6.101}$$

Example 6.11 Find the velocity of a raindrop of mass $m(t)$ if the rate at which the droplet accumulates mass is constant. Assume that $v(0) = 0$ and $m(0) = m_0$.

Solution The droplet accumulates mass; therefore,

$$\frac{dm(t)}{dt} = k, \quad k > 0.$$

From the initial conditions

$$m(t) = m_0 + kt.$$

Substituting this into (6.101) and integrating, remembering to use $v(0) = 0$, yields

$$v(t) = \frac{1}{2}gt\left(2 + \frac{kt}{m_0}\right)\left(1 + \frac{kt}{m_0}\right)^{-1}. \quad \square$$

6.8 Exercises

1. A 3-particle system has particles of mass 1, 2 and 3 positioned at points $(1,1,1)$, $(1,-2,0)$ and $(0,2,1)$, respectively. Find the position of the centre of mass.

2. An asteroid after entering the Earth's atmosphere breaks up into constituent parts. Show that the locus of the centre of mass is part of a parabola. (Do not consider air resistance.) [Hint: The path of the projectile moving freely in a uniform gravitational field is parabolic.]

   Explain why (or why not) the angular momentum is conserved.

3. A 4-particle system has particles of mass 1, 2, 3 and 1 positioned at points $(2t - 3, 2t^2, t^3)$, $(-3, 8 - 3t^3, -4t)$, $(2 - t, t^{-1}, \ln t)$ and $(-22, 43, t)$, respectively. Time is represented by $t \neq 0$. Find the velocity of the centre of mass, the total linear momentum and the total angular momentum about the origin.

4. Starting with equation (6.1), show that the total work done, $T(t_1) - T(t_2)$, in moving an N-particle system from an initial configuration at time t to a final configuration at time t_2, is given by

$$T(t_2) - T(t_1) = \sum_{i=1}^{N} \int_{t_1}^{t_2} \mathbf{F}^{\text{ext}} \cdot \dot{\mathbf{r}}_i dt + \sum_{i=1}^{N}\sum_{j\neq i}^{N} \int_{t_1}^{t_2} \mathbf{F}_{ji} \cdot \dot{\mathbf{r}}_i dt.$$

   [Hint: $\dot{\mathbf{r}}_i \cdot \ddot{\mathbf{r}}_i \equiv \frac{1}{2}d\dot{\mathbf{r}}_i^2/dt$.]

   A 3-particle system has particles of unit mass. The system is subject to a force such that the position vectors of the particles, with respect to a fixed origin, are $\mathbf{r}_1 = (t, t^2 - 2, 3)$, $\mathbf{r}_2 = (\ln t^2, -t^2, 1)$ and $\mathbf{r}_3 = (t - 1, 3 - t, t)$. Time is represented by $t \neq 0$. Find the total work done by the force in moving the 3-particle system from an initial configuration at time $t = 1$ to a final configuration at time $t = 2$.

5. Two smooth balls collide with each other, one of the balls being at rest. Assuming that the collision is elastic, show that the masses of the balls are the same provided that the scattering angle and recoil angle sum to $\pi/2$.

6. Two smooth balls collide with each other. Let the mass of the balls be m_1 and m_2. Their initial velocities are the same, but after the collision the velocity of m_2 is zero. What is the mass of m_1 in terms of m_2?

7. Three smooth balls B_1, B_2 and B_3 of mass m, m and $2m$, respectively, are restricted to travel only along a horizontal line that passes through their centres of mass (see Figure 6.14). Initially, B_2 and B_3

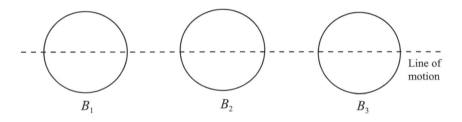

FIGURE 6.14
Three particles in collision

are at rest, whilst B_1 is travelling towards B_2 with speed u. After the final collision has occurred, find the velocities of the three balls.

Suppose B_2 is now removed so that B travels towards B_3 at velocity u and B_3 is at rest. What are the velocities of B_1 and B_3 after the collision? Comment on your results.

8. Two smooth particles having masses of m_1 and m_2 collide elastically and head-on. Before they collide the velocities of m_1 and m_2 are u_1 and u_2, respectively; following the collision they are v_1 and v_2, respectively. Show that these velocities satisfy the relation

$$u_1 - u_2 = v_2 - v_1.$$

9. N particles $P_1, P_2, \ldots, P_N$ with identical masses are restricted to

travel only along a horizontal line (see Figure 6.15). The particles collide elastically. If the initial speeds of the particles are such that

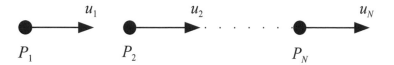

FIGURE 6.15

N particles in collision

$u_1 > u_2 > \cdots > u_N$ what are the final speeds of the particles after the collision?

10. A ball of mass m_1 is suspended fractionally above a ball of mass m_2, which is in turn suspended fractionally above a ball of mass m_3. The masses are such that $m_1 \leq m_2 \leq m_3$. They are suspended such that a line joining their centres of mass is precisely vertical (see Figure 6.16). The balls are then released simultaneously. After all the collisions have occurred, the ball of mass m_1 reaches a height of 500 metres. What is the approximate value of h, the height between the ground and the ball of mass m_3?

11. A pair of particles P_1 and P_2 with masses m_1 and m_2 collide inelastically in the LAB frame. P_2 is at rest. The collision produces a pair of particles with masses m_3 and m_4. These particles travel away at angles α and β with respect to the original path of P_1. Show that the energy absorbed during the collision is

$$Q = \frac{P_1^2}{2m_1}\left[1 - \frac{(m_1/m_3)\sin^2\beta + (m_1/m_4)\sin^2\alpha}{\sin^2(\alpha+\beta)}\right].$$

12. A ball is released from a height h above level ground. If the coefficient of restitution is e, show that the time taken for the ball to come to rest is

$$\sqrt{\frac{2h}{g}}\left(\frac{1+e}{1-e}\right),$$

and the total distance covered by the ball is

$$h\left(\frac{1+e^2}{1-e^2}\right).$$

13. A puck slides smoothly on ice and collides with a fixed vertical

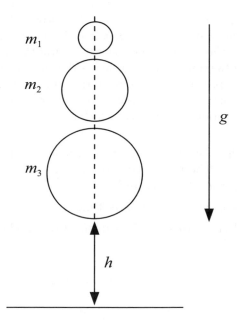

m_1

m_2

g

m_3

h

FIGURE 6.16
Three balls colliding vertically

hockey stick blade. If the puck collides with the blade at an angle α and speed u, show that it leaves the blade at an angle

$$\tan^{-1}(e \tan \theta)$$

and speed

$$u\sqrt{e^2 + (1 - e^2)\cos^2 \theta}.$$

14. Three particles P_1, P_2 and P_3 lie in a straight line; they have masses $2, 3$ and 4, respectively. Initially, P_2 and P_3 are at rest and P_1 travels towards P_2 (head-on) at speed $2u$. After the collision, P_2 travels towards P_3 (head-on) at speed u, colliding with P_3 and subsequently coming to rest. Find the coefficients of restitution between P_1 and P_2, and between P_2 and P_3. What are the final speeds of P_1, P_2 and P_3? Leave your answer in terms of u.

15. Find the velocity of a raindrop as it falls from rest through a stationary cloud if its mass increases at a rate

$$\frac{dm(t)}{dt} = km(t).$$

16. A green racing car and a red racing car compete in a race along a straight road. The driver plus car has mass m_0. The driver of the green car has been informed by his team that it will rain at some point during the race. So the team decide cleverly to drill the floor of the green car to allow the rain to drain away — thus not adding to the mass of the car. As predicted, it begins to rain at some point during the race. This occurs, bizarrely, when both cars run out of fuel and enter a smooth section of the track, which remains smooth until the end of the race.

The rainwater entering the green car drains away without contributing to the mass of the car. The rainwater entering the red car increases its mass by $m(t) = m_0 + kt$, where $k > 0$.

Determine the velocities of each car at a later time t.

The race is won by the last car to stop. Was green's team as clever as they thought?

7

Rigid Bodies

CONTENTS

A *rigid body* is defined as a system of N particles for which the distance between any two particles is constant. Thus, for two arbitrarily chosen particles in the body P_i and P_j, we have

$$|\mathbf{r}_i - \mathbf{r}_j| = \text{constant.} \tag{7.1}$$

As is the case for an N-particle system, a rigid body's motion depends on the motion of the centre of mass and the motion relative to the centre of mass — the latter motion, necessarily on account of (7.1), being a rotation.

7.1 Rotation of a rigid body about a fixed axis

Our discussion will solely concern aspects of a rigid body's motion about a fixed axis. In Section 7.2, we will turn our attention to a rigid body's combined rotational and translational motion.

7.1.1 Angular velocity and kinetic energy

As will shortly become evident, the *moment of inertia* about and axis of a rigid body is the rotational analogue of mass in linear motion. Indeed, as the theory of rigid bodies develops, the concept of mass that appears in the definitions of angular momentum and kinetic energy will be replaced by the moment of inertia.

Consider a particle P in $\mathbb{E}^3$, initially located at A, that is free to rotate about a fixed axis, where $\hat{\mathbf{n}}$ is a unit vector along the axis (see Figure 7.1). The perpendicular distance between the particle and the axis is s. At A, the

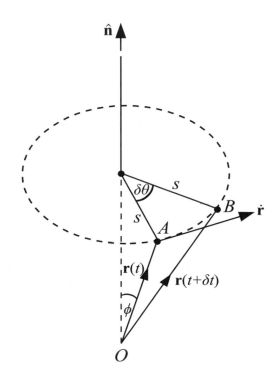

FIGURE 7.1
Particle rotating about a fixed axis

position of the particle relative to O is $\mathbf{r}(t)$. A short time δt later, the particle moves through a small angle $\delta\theta$ to the point B with position vector $\mathbf{r}(t + \delta t)$. The velocity of the particle at A is, therefore,

$$\dot{\mathbf{r}} = \lim_{\delta t \to 0} \frac{\mathbf{r}(t + \delta t) - \mathbf{r}(t)}{\delta t}.$$

It can be seen from Figure 7.1 that $\dot{\mathbf{r}}$ is always perpendicular to $\mathbf{r}$ and also perpendicular to the plane spanned by $\hat{\mathbf{n}}$ and $\mathbf{r}$. Hence,

$$(\hat{\mathbf{n}} \times \mathbf{r})\lambda(t) = \dot{\mathbf{r}},$$

where $\lambda(t)$ is a scalar. However,

$$\left|\frac{d\mathbf{r}}{dt}\right| = s\frac{d\theta}{dt} = |\mathbf{r}|\sin\phi\frac{d\theta}{dt} = |\hat{\mathbf{n}} \times \mathbf{r}|\frac{d\theta}{dt}, \qquad (7.2)$$

where $d\theta/dt = \lim_{\delta t \to 0} \delta\theta/\delta t$. The inference is that $\lambda(t) = d\theta/dt = \dot{\theta}$. Therefore,

$$\dot{\theta}\hat{\mathbf{n}} \times \mathbf{r} = \dot{\mathbf{r}}.$$

Introducing the quantity $\boldsymbol{\omega} = \omega\hat{\mathbf{n}} = \dot{\theta}\hat{\mathbf{n}}$, we have

$$\boldsymbol{\omega} \times \mathbf{r} = \dot{\mathbf{r}}, \qquad (7.3)$$

where $\boldsymbol{\omega}$ is the *angular velocity* of the particle and $\omega = |\boldsymbol{\omega}|$ is the angular speed, met previously.

Notice that because $\hat{\mathbf{n}}$ points in a direction perpendicular to the plane of rotation, then so will $\boldsymbol{\omega}$. A simple way of remembering the direction of the angular velocity is to imagine removing a screw from a block of wood. On turning the screw anticlockwise (representing the rotation of the particle) the screw begins to rise from the block of wood (representing the *positive* direction of the angular velocity).

Next, we shall see how the kinetic energy of the particle rotating around the fixed axis can be interpreted in terms of the angular velocity. By definition, the kinetic energy for a single particle is

$$T = \frac{1}{2}m|\dot{\mathbf{r}}|^2,$$

where m is the mass of the particle. Using (7.2) and (7.3), this yields

$$T = \frac{1}{2}m|\boldsymbol{\omega} \times \mathbf{r}|^2 = \frac{1}{2}m|\hat{\mathbf{n}} \times \mathbf{r}|^2\dot{\theta}^2 = \frac{1}{2}ms^2\omega^2. \qquad (7.4)$$

7.1.2 Moment of inertia

The discussion in the previous section concerned only the motion of a single particle. Extending the notions of angular velocity and kinetic energy to a rigid body is straightforward.

A property of a rigid body is that every constituent particle of the rigid body must have the same value of angular velocity. Let us consider the ith particle of the rigid body. Writing (7.3) for the ith particle

$$\boldsymbol{\omega} \times \mathbf{r}_i = \dot{\mathbf{r}}_i. \qquad (7.5)$$

Differentiating (7.1):

$$\frac{d}{dt}|\mathbf{r}_i - \mathbf{r}_j|^2 = 2(\mathbf{r}_i - \mathbf{r}_j) \cdot (\dot{\mathbf{r}}_i - \dot{\mathbf{r}}_j)$$

$$= 2(\mathbf{r}_i - \mathbf{r}_j) \cdot \boldsymbol{\omega} \times (\mathbf{r}_i - \mathbf{r}_j)$$

$$= 0.$$

So, (7.1) holds provided $\boldsymbol{\omega}$ is the same for all constituent particles of a rigid body.

From the definition of the kinetic energy of a multi-particle system (6.20):

$$T = \sum_{i=1}^{N} \frac{1}{2} m_i |\dot{\mathbf{r}}_i|^2 = \sum_{i=1}^{N} \frac{1}{2} m_i |\boldsymbol{\omega} \times \mathbf{r}_i|^2 = \sum_{i=1}^{N} \frac{1}{2} m_i s_i^2 \omega^2$$

or

$$T = \frac{1}{2} I \omega^2, \qquad (7.6)$$

where

$$I = \sum_{i=1}^{N} m_i s_i^2 \qquad (7.7)$$

is the *moment of inertia* of a rigid body about a specified axis — in this case, the axis was aligned in the direction of the unit vector $\hat{\mathbf{n}}$.

Comparing (7.6) with the kinetic energy of a single particle performing linear motion ($T = \frac{1}{2} m v^2$) we can clearly see the analogy between m and I, and v and ω. In (7.6), T is called the *rotational kinetic energy*.

7.1.3   Angular momentum

From the definition of the angular momentum of a multi-particle system (6.9)

$$\mathbf{L} = \sum_{i=1}^{N} m_i \mathbf{r}_i \times \dot{\mathbf{r}}_i = \sum_{i=1}^{N} m_i \mathbf{r}_i \times (\boldsymbol{\omega} \times \mathbf{r}_i) = \sum_{i=1}^{N} m_i \mathbf{r}_i \times (\omega \hat{\mathbf{n}} \times \mathbf{r}_i).$$

Taking the scalar product of this expression with the unit vector $\hat{\mathbf{n}}$ yields

$$\mathbf{L} \cdot \hat{\mathbf{n}} = \sum_{i=1}^{N} m_i \omega [\mathbf{r}_i \times (\hat{\mathbf{n}} \times \mathbf{r}_i)] \cdot \hat{\mathbf{n}}$$

$$= \sum_{i=1}^{N} m_i \omega (\hat{\mathbf{n}} \times \mathbf{r}_i) \cdot (\hat{\mathbf{n}} \times \mathbf{r}_i)$$

$$= \sum_{i=1}^{N} m_i \omega |\hat{\mathbf{n}} \times \mathbf{r}_i|^2$$

$$= \sum_{i=1}^{N} m_i s_i^2 \omega$$

$$= I \omega. \qquad (7.8)$$

The second line arises due to the fact that vectors forming a scalar triple product can be cyclically permutated.

Equation (7.8) gives the component of the angular momentum in the directions along the axis of rotation; namely, $\hat{n}$. However, in general, the fixed axis of rotation is not parallel with $\mathbf{L}$. So

$$|\mathbf{L}|\cos\theta = I\omega,$$

where θ is the angle between $\mathbf{L}$ and $\hat{n}$. If $\mathbf{L}$ and $\hat{n}$ are parallel then the angular momentum takes the simple form

$$\mathbf{L} = I\omega\hat{n}. \tag{7.9}$$

7.1.4 Calculating the centre of mass and the moment of inertia of some uniform rigid bodies

A rigid body is an N-particle system, subject to (7.1), where N can be extremely large indeed. It is, therefore, more convenient to consider a rigid body as a continuous distribution of matter of density $\rho = dM/dV$, where dM is the total element of mass within a volume dV. The mass element dM is chosen so that the density distribution is constant, otherwise ρ will be position dependent and the rigid body will not be uniform.

The discrete form of the total mass (6.2) can now be replaced by the volume integral

$$M = \iiint_V \rho dV. \tag{7.10}$$

Similarly, the discrete form of the position of the centre of mass can be replaced by the volume integral

$$R = \frac{1}{M} \iiint_V \rho \mathbf{r} dV$$

or, because the density distribution is constant,

$$R = \frac{1}{V} \iiint_V \mathbf{r} dV. \tag{7.11}$$

Although we will, invariably, have to resort to (7.11) when determining the centre of mass of all except the most symmetrical of rigid bodies, a little astute observation with regard to the symmetry properties of some rigid bodies can give a good indication of the approximate position of the centre of mass. For example, consider the hyperboloid of two sheets in Figure 7.2. It can be seen that it has *reflectional symmetry* in the yz-plane, which means that the centre of mass must be somewhere in this plane. Moreover, this rigid body has *rotational symmetry* about the x-axis, which means that the centre of mass must also be somewhere along this axis. The point where the x-axis intersects the yz-plane must, therefore, be the precise location of the centre of mass, namely the origin. This example also illustrates that the centre of mass of a rigid body need not always be located somewhere within the body.

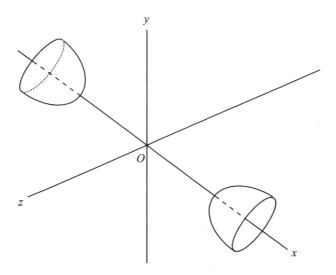

FIGURE 7.2
Hyperboloid of two sheets

If one considers a cone, say, it is quite easy without calculation to determine a region in which the centre of mass will be located. The cone has rotational symmetry about an axis that passes through its vertex and is perpendicular to the base of the cone. Therefore, the centre of mass must be located somewhere along this axis, but to find the precise location a detailed calculation must be employed.

Example 7.1 Calculate the position of the centre of mass of a uniform solid hemisphere of radius l.

Solution The hemisphere is depicted in Figure 7.3. It is clear without embarking on any formal calculation that the centre of mass is located somewhere along the z-axis due to rotational symmetry.
    To tackle this problem, we will transform to spherical polar coordinates

$$x = r \sin \theta \cos \phi$$
$$y = r \sin \theta \sin \phi$$
$$z = r \cos \theta$$

and employ (7.11). The volume element becomes $dV = r^2 \sin \theta dr d\theta d\phi$, where the volume of the hemisphere is $V = 2\pi l^3/3$. Hence,

$$\mathbf{R} = \frac{3}{2\pi l^3} \int_{r=0}^{l} \int_{\theta=0}^{\pi/2} \int_{\phi=0}^{2\pi} (x, y, z) r^2 \sin \theta dr d\theta d\phi.$$

However, it has already been ascertained that the centre of mass is

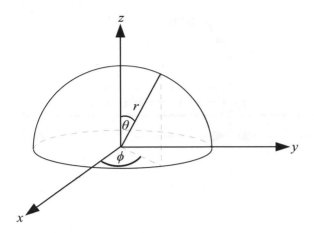

FIGURE 7.3
Centre of mass of a solid hemisphere

located along the z-axis, implying that the coordinates $x = 0 = y$. Thus writing $\mathbf{R} = (X, Y, Z)$, we have

$$Z = \frac{3}{2\pi l^3} \int_{r=0}^{l} \int_{\theta=0}^{\pi/2} \int_{\phi=0}^{2\pi} r^3 \cos\theta \sin\theta \, dr \, d\theta \, d\phi$$

$$= \frac{3}{2\pi l^3} \left(\frac{l^4}{4}\right)(2\pi) \int_{\theta=0}^{\pi/2} \frac{1}{2}\sin(2\theta)d\theta = \frac{3}{8}l.$$

Hence, the position of the centre of mass is $\mathbf{R} = \frac{3}{8}(0, 0, l)$. $\qquad \square$

The moment of inertia (7.7) can be similarly expressed as an integral over the volume

$$I = \iiint_V \rho s^2 dV \tag{7.12}$$

or in terms of spherical polar coordinates (see Figure 7.1)

$$I = \iiint_V \rho (r\sin\theta)^2 dV. \tag{7.13}$$

We will now consider the problem of calculating the moment of inertia for some simple uniform rigid bodies. One important point to remember when calculating moments of inertia is that it is necessary to specify the axis about which the rotation is taking place. For example, the moment of inertia of a thin rod about a perpendicular axis through its centre is, of course, different from that taken at one end.

Thin rod Consider a thin rod of mass M and length $2l$ as depicted in Figure 7.4. We want to find the moment of inertia of the rod about a perpendicular

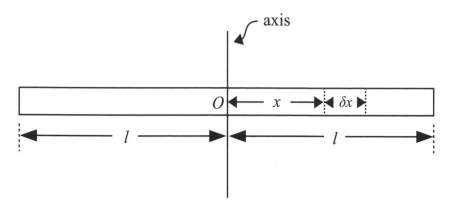

FIGURE 7.4
Thin rod with an axis through its centre

axis through its centre of mass. This is a one-dimensional problem; hence, the density of the rod is given as a mass per unit length, so in this case $\rho = M/2l$.

The perpendicular distance of the element δx from the axis is x. The total one-dimensional 'volume' of the rod is $2l$. Employing (7.12) and replacing s with x and dV with dx, we find, on integrating over $-l \leq x \leq l$, the moment of inertia to be

$$I^c = \int_{-l}^{l} \rho x^2 dx = \rho \left[\frac{x^3}{3} \right]_{-l}^{l} = \frac{2}{3}\rho l^3 = \frac{1}{3}Ml^2. \qquad (7.14)$$

Consider the same rod only with the perpendicular axis at one end. Nothing changes in our discussion above except for the limits of integration; they become $0 \leq x \leq 2l$. We find on repeating the calculation that the moment of inertia is

$$I = \frac{8}{3}\rho l^3 = \frac{4}{3}Ml^2. \qquad (7.15)$$

Thin circular ring Consider a thin circular ring of mass M and radius l as depicted in Figure 7.5. We want to find the moment of inertia of the ring about an axis that passes through the centre of mass and perpendicular to the plane of the ring. Think of the ring as being composed of N individual masses $m_1, m_2, \ldots, m_N$. The distance between each mass m_i and the axis is l. We can, therefore, employ (7.7) and replace s_i with l. Hence

$$I^c = \sum_{i=1}^{N} m_i l^2 = Ml^2. \qquad (7.16)$$

Thin circular disc Consider a thin circular disc of mass M and radius l as depicted in Figure 7.6. We want to find the moment of inertia of the disc

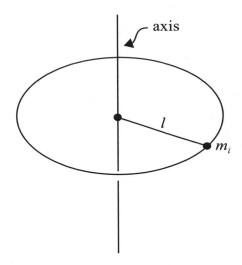

FIGURE 7.5
Thin circular ring with an axis through its centre

about an axis through the centre of mass and perpendicular to the plane of the disc. This is a two-dimensional problem; hence, the density of the disc is given as a mass per unit area, so in this case $\rho = M/\pi l^2$.

The perpendicular distance of the element δr from the axis is r. The total two-dimensional 'volume' of the disc is πl^2. So as to cover the entire region of the disc, we must integrate over r and θ. The triple integral (7.12) then becomes the double integral

$$I = \iint_A \rho s^2 dA,$$

where A is the entire two-dimensional region occupied by the disc. On replacing s with r and dA with $r dr d\theta$ (incremental area in plane polar coordinates), we find on incorporating the ranges of integration $0 \leq r \leq l$ and $0 \leq \theta \leq 2\pi$ that the moment of inertia is

$$I^c = \int_0^l \int_0^{2\pi} \rho r^2 r \, dr \, d\theta$$

$$= \rho \left(\int_0^l r^3 dr \right) \left(\int_0^{2\pi} d\theta \right)$$

$$= \rho \left(\frac{l^4}{4} \right) (2\pi)$$

$$= \frac{1}{2} \rho \pi l^4 = \frac{1}{2} M l^2. \tag{7.17}$$

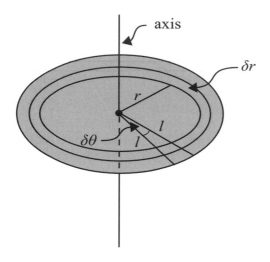

FIGURE 7.6
Thin circular disc with an axis through its centre

Consider the same disc only with the axis through the centre of mass and lying in the plane of the disc as depicted in Figure 7.7. The axis is orientated so that it coincides with $\theta = 0$. Each point on the disc is a perpendicular distance $s = r \sin \theta$ from the axis of rotation. Therefore, the moment of inertia is

$$
\begin{aligned}
I^c &= \int_0^l \int_0^{2\pi} \rho(r \sin \theta)^2 r dr d\theta \\
&= \frac{1}{2}\rho\left(\int_0^l r^3 dr\right)\left(\int_0^{2\pi} (1 - \cos 2\theta)d\theta\right) \\
&= \frac{1}{2}\rho\left(\frac{l^4}{4}\right)(2\pi) \\
&= \frac{1}{4}\rho\pi l^4 = \frac{1}{4}Ml^2.
\end{aligned}
\tag{7.18}
$$

Solid cylinder Consider a right circular cylinder of mass M, radius l and length h as depicted in Figure 7.8. Notice this is merely the disc in Figure 7.6 extended down a distance h. This is, of course, a three-dimensional problem; hence, the density of the disc is given as a mass per unit volume, so in this case $\rho = M/\pi l^2 h$.

This problem is most suited to cylindrical polar coordinates. Thence, s is replaced by r and dV is replaced by $r dr d\theta dz$ (incremental volume in cylindrical polar coordinates) in (7.12). The range of integration is $0 \le r \le l, 0 \le \theta < 2\pi$

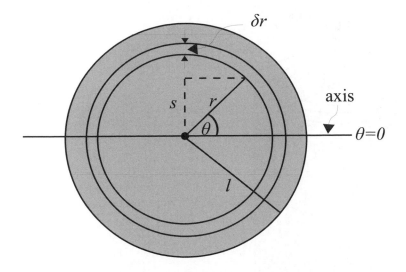

FIGURE 7.7
Thin circular disc with an axis lying in the plane of the disc

and $0 \leq z \leq h$. The moment of inertia is

$$I^c = \int_0^r \int_0^{2\pi} \int_0^h \rho r^2 r \, dr \, d\theta \, dz$$

$$= \rho \left(\frac{l^4}{4} \right)(2\pi)(h)$$

$$= \frac{1}{2}\rho\pi l^4 h = \frac{1}{2}Ml^2. \tag{7.19}$$

Solid sphere Consider a solid sphere of mass M and radius l as depicted in Figure 7.9. We want to find the moment of inertia of the sphere about an axis through its centre of mass which is orientated so that it coincides with $\theta = 0$. The density of the sphere is $\rho = M/(4/3)\pi l^3$. Each point in the sphere is a perpendicular distance $s = r \sin \theta$ from the axis of rotation.

This problem is most suited to spherical polar coordinates. Thence, s is replaced by $r \sin \theta$ and dV is replaced by $r^2 \sin \theta \, dr \, d\theta \, d\phi$ (incremental volume in spherical polar coordinates) in (7.12). The range of integration is $0 \leq r \leq l$, $0 \leq \theta < \pi$ and $0 \leq \phi < 2\pi$. The moment of inertia is

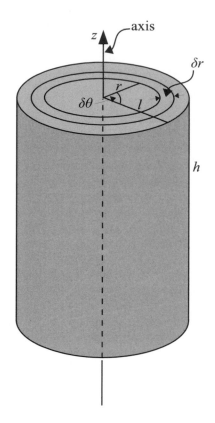

FIGURE 7.8
Solid cylinder with an axis through its centre

$$I^c = \int_0^l \int_0^\pi \int_0^{2\pi} \rho(r\sin\theta)^2 r^2 \sin\theta \, dr \, d\theta \, d\phi$$

$$= \rho\left(\int_0^l r^4 dr\right)\left(\int_0^\pi \sin^3\theta \, d\theta\right)\left(\int_0^{2\pi} d\phi\right)$$

$$= \rho\left(\frac{l^5}{5}\right)\left(\frac{4}{3}\right)(2\pi)$$

$$= \frac{8}{15}\rho\pi l^5 = \frac{2}{5}Ml^2. \tag{7.20}$$

The semicircular lamina Consider a thin semicircular lamina[1] of mass M and radius l. This is merely Figure 7.7 with the half under the axis of rotation removed. We want to find the moment of inertia of the lamina about the axis $\theta = 0$. The only changes to the original calculation of the circular disc are the

---

[1]This is a generic term for any two-dimensional plane surface.

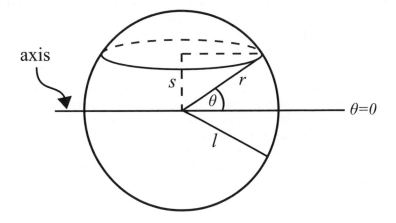

FIGURE 7.9
Solid sphere with an axis through its centre

area $(\pi l^2 / 2)$ and range $(0 \leq \theta \leq \pi)$. The density of the lamina is $\rho = 2M/\pi l^2$, so the moment of inertia is

$$I = \int_0^l \int_0^{2\pi} \rho (r \sin \theta)^2 r \, dr \, d\theta$$

$$= \frac{1}{8} \rho \pi l^4 = \frac{1}{4} M l^2. \qquad (7.21)$$

It may not have escaped the reader's notice that in two cases, the moment of inertia of two rigid bodies is the same for appropriately chosen axes of rotation.

Firstly, consider the circular disc and the solid cylinder; both have their axes of rotation along their lines of symmetry. If these bodies are viewed from a particular perspective so that the axis of each body lies parallel to the direct line of vision, both the disc and cylinder will appear as identical two-dimensional circular laminae. This is true for any two- and three- dimensional bodies that can be viewed in this way. For example, the moments of inertia will be the same for a solid rectangular lamina of mass M and sides $2a$ and $2b$ about an axis through its centre and perpendicular to its plane, and a solid block of mass M sides $2a$, $2b$ and depth $2c$ about an axis through its centre and perpendicular to the plane of area $4ab$. In both cases the moment of inertia is $1/3M(a^2 + b^2)$, *but* M in each case will, of course, be different. Note, however, this would not apply to, say, a circular disc and a solid cone when viewed in this manner. Unlike the cylinder, the cone tapers off to a point and so the perpendicular distance between points on the surface of the cone and the axis will vary between base and apex. This will affect the value of the moment of inertia. Furthermore, the above discussion is also true for one- and two-dimensional bodies that can be viewed in like manner. For example, the thin rod of mass M and length $2a$ and perpendicular axis through its centre will have the same moment of inertia as the solid rectangular lamina with length $2a$, depth $2b$ and axis through its centre and parallel to the plane.

Next, consider the circular disc in Figure 7.7 and the semi-circular lamina. Removing the bottom (or top) half segment of the disc to produce the semi-circular lamina will yield the same value for the moment of inertia in each case provided that the axis is not reorientated. This is true for any body that allows segments to be removed in this way. For example, the moment of inertia of a solid sphere about an axis through its centre will be the same if segments of the sphere had been removed (like breaking up an orange) leaving just one segment 'attached' to the axis. Again, the masses will be different in each case.

7.1.5   Parallel axis theorem

Suppose the moment of inertia of a rigid body has been determined about an axis through its centre of mass — it does not matter which axis is taken provided that it passes through the body's centre of mass. It is then possible to deduce the moment of inertia of the body about any other *parallel* axis. Let us state this as a theorem.

Theorem 1 *Given the moment of inertia I^c of a rigid body about an axis through its centre of mass, then the moment of inertia I about a parallel axis is*

$$I = I^c + Md^2, \tag{7.22}$$

where M is the mass of the body and d is the distance between the axes.

Proof 1 *Consider the lamina in Figure 7.10. Two axes have been drawn: the centre of mass axis (solid line) that passes through the centre of mass C and a general axis (dashed line) parallel to the centre of mass axis and displaced a distance d. Both axes are aligned in a direction specified by the unit vector $\hat{\mathbf{n}}$. A particle P_i of mass m_i has position vector r_i relative to an origin O that lies somewhere along the general axis, and position vector r_i^c relative to the centre of mass C.*

From Figure 7.10, we have $\mathbf{r}_i = \mathbf{R} + \mathbf{r}_i^c$. Then $d = |\hat{\mathbf{n}} \times \mathbf{R}|$, $d_i^c = |\hat{\mathbf{n}} \times \mathbf{r}_i^c|$ and the perpendicular distance of P_i from the general axis is $s_i = |\hat{\mathbf{n}} \times \mathbf{r}_i|$.

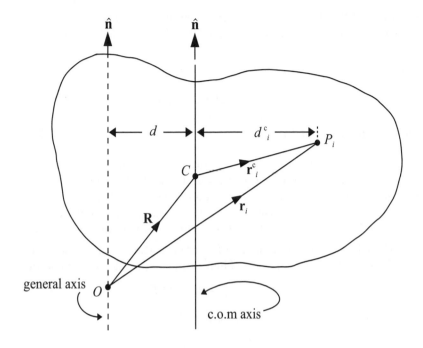

FIGURE 7.10
Parallel axis theorem

Now, the moment of inertia of the lamina about the general axis is

$$
I = \sum_{i=1}^{N} m_i s_i^2 = \sum_{i=1}^{N} m_i |\hat{\mathbf{n}} \times \mathbf{r}_i|^2 = \sum_{i=1}^{N} m_i |\hat{\mathbf{n}} \times (\mathbf{R} + \mathbf{r}_i^c)|^2
$$

$$
= \sum_{i=1}^{N} m_i (\hat{\mathbf{n}} \times \mathbf{R} + \hat{\mathbf{n}} \times \mathbf{r}_i^c) \cdot (\hat{\mathbf{n}} \times \mathbf{R} + \hat{\mathbf{n}} \times \mathbf{r}_i^c)
$$

$$
= \sum_{i=1}^{N} m_i |\hat{\mathbf{n}} \times \mathbf{R}|^2 + \sum_{i=1}^{N} m_i |\hat{\mathbf{n}} \times \mathbf{r}_i^c|^2 + 2 \sum_{i=1}^{N} m_i (\hat{\mathbf{n}} \times \mathbf{R}) \cdot (\hat{\mathbf{n}} \times \mathbf{r}_i^c)
$$

$$
= \sum_{i=1}^{N} m_i d^2 + \sum_{i=1}^{N} m_i (d_i^c)^2 + 2(\hat{\mathbf{n}} \times \mathbf{R}) \cdot (\hat{\mathbf{n}} \times \sum_{i=1}^{N} m_i \mathbf{r}_i^c).
$$

The first sum is merely Md^2. The second sum is the moment of inertia about the centre of mass axis. The third term has $\sum_{i=1}^{N} m_i \mathbf{r}_i^c$ as part of the vector product; however, as we know from (6.14) this must vanish (the position of the centre of mass relative to itself is zero). Thence, equation (7.22) follows.

Example 7.2 Using the parallel axis theorem, calculate the moment of inertia of a uniform thin rod of mass M and length $2l$ about a perpendicular axis at one end.

Solution The moment of inertia $I^c = Ml^2/3$ by (7.14). Hence,

$$I = \frac{1}{3}Ml^2 + Ml^2 = \frac{4}{3}Ml^2,$$

which is the result (7.15).   □

Example 7.3 A uniform lamina of mass M has three parallel axes lying in its plane, with one axis passing through the centre of mass and the other two axes on opposite sides of the axis. If the two axes are a distance d_1 and d_2 from the centre of mass axis, calculate the difference in the moments of inertia of the lamina about these axes.

Solution Let I^c be the moment of inertia of the lamina about the axis through the centre of mass. Then from the parallel axis theorem

$$I^c = I_1 - Md_1^2 = I_2 - Md_2^2.$$

So the difference in the moments of inertia is

$$I_1 - I_2 = M(d_1^2 - d_2^2).$$ □

7.1.6   Perpendicular axis theorem

Suppose the moment of inertia of a rigid body has been determined about two perpendicular axes. Then it is possible to deduce the moment of inertia of the body about an axis that is mutually perpendicular to the other two axes. Let us state this as a theorem.

Theorem 2 *Given that the moments of inertia of a lamina about perpendicular axes (Ox and Oy lying in the plane of the lamina) are I_x and I_y, respectively, the moment of inertia I_z about an axis Oz perpendicular both to Ox and Oy is given by*

$$I_z = I_x + I_y. \tag{7.23}$$

Proof 2 *Set up a three-dimensional Cartesian frame such that the axes Ox and Oy lie in the lamina and Oz is perpendicular to the lamina (Figure 7.11). A constituent particle of mass m_i is located a perpendicular distance x_i from the y-axis, y_i from the x-axis and z_i from the z-axis. Taking the sum of the moment of inertia I_x and I_y yields*

$$I_x + I_y = \sum_{i=1}^{N} m_i y_i^2 + \sum_{i=1}^{N} m_i x_i^2 = \sum_{i=1}^{N} m_i(y_i^2 + x_i^2) = \sum_{i=1}^{N} m_i z_i^2 = I_z.$$

This completes the proof.

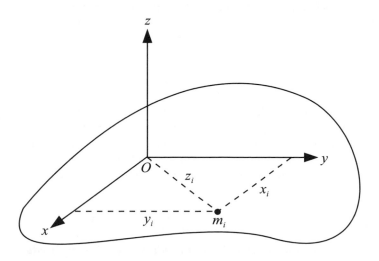

FIGURE 7.11
Perpendicular axis theorem

Example 7.4 Using the perpendicular axis theorem, calculate the moment of inertia of a uniform solid disc of mass M and radius l about an axis through the centre of mass and perpendicular to the disc.

Solution The moment of inertia of such a disc was already calculated about an axis along its diameter. This was, by result (7.18), $Ml^2/4$. Due to the symmetry of the disc, this result holds true about any diameter of the disc. Thus, if Ox and Oy lie in the plane of the disc, we have from (7.23)

$$I_z = \frac{1}{4}Ml^2 + \frac{1}{4}Ml^2 = \frac{1}{2}Ml^2.$$

Note this result was derived using integration (see 7.17). □

Example 7.5 Calculate the moment of inertia for a solid cube of mass M and edge $2l$ about an axis along any edge.

Solution We begin by considering the rod in Figure 7.4. The moment of inertia was found to be

$$I_{\text{rod}} = \frac{1}{3}M_{\text{rod}}l^2$$

about an axis through its centre. Now extend the rod so that it becomes a solid square lamina, each edge measuring $2l$. The moment of inertia of this lamina about an axis through its centre and parallel to the plane of the lamina must be the same as that of the rod (see discussion above). Only the mass of the rod is replaced with the mass of the lamina M_{lamina}:

$$I_{\text{lamina}} = \frac{1}{3} M_{\text{lamina}} l^2.$$

Now rotate the axis by $\pi/2$ about the centre of the lamina so that it still lies in the plane of the lamina but it is passing through an adjacent perpendicular edge. By symmetry, the moment of inertia of the lamina about the axis is also $M_{\text{lamina}} l^2/3$. So using the perpendicular axis theorem, the moment of inertia perpendicular to the lamina and passing through its centre is

$$I_{\text{lamina/perpendicular}} = \frac{1}{3} M_{\text{lamina}} l^2 + \frac{1}{3} M_{\text{lamina}} l^2 = \frac{2}{3} M_{\text{lamina}} l^2.$$

Now extend the lamina so that it becomes a solid cube, each edge measuring $2l$. The moment of inertia of the cube about an axis through its centre of mass and perpendicular to a face must be the same as that of the lamina (see discussion above). Hence,

$$I_{\text{cube/c.o.m.}} = \frac{2}{3} M_{\text{cube}} l^2,$$

where M_{cube} is the mass of the cube. Using the parallel axis theorem, the moment of inertia of the cube about an axis along one edge is

$$I_{\text{cube/edge}} = \frac{2}{3} M_{\text{cube}} l^2 + M_{\text{cube}} (\sqrt{2} l)^2$$
$$= \frac{8}{3} M_{\text{cube}} l^2. \qquad \square$$

7.2 Planar motion of a rigid body

Planar motion is exactly what it sounds like: motion restricted to a plane. However, the rigid body performing the motion need not be a lamina; any three-dimensional body can perform planar motion provided that the motion is restricted along two axes, say the x- and y-axis, and, therefore, no motion takes place along the z-axis. Examples of rigid bodies performing planar motion are (i) a rod oscillating about an axis through one end, and (ii) a cylinder rolling down an incline. However, a planet orbiting a star is not, strictly speaking, planar motion; this is because the planet will not, in general, be rotating about an axis perpendicular to the orbital planar motion. Notice that the planar motion in (i) is purely rotational, while that in (ii) combines rotational and translational motion.

7.2.1 Kinetic energy

Recall that the total kinetic energy of a multi-particle system is given by (6.21), namely,

$$T = \sum_{i=1}^{N} \frac{1}{2} m_i \dot{\mathbf{R}}^2 + \sum_{i=1}^{N} \frac{1}{2} m_i (\mathbf{r}_i^c)^2$$

$$= \frac{1}{2} M \dot{\mathbf{R}}^2 + T^c,$$

where T^c is the kinetic energy of the system relative to its centre of mass, M is the total mass of the system and $\dot{\mathbf{R}}$ is the velocity of the centre of mass. When the system is a rigid body, the term $(1/2)M\dot{\mathbf{R}}^2$ is referred to as the *translational kinetic energy*, whilst T^c is referred to as the *rotational kinetic energy*. To see the form that T^c takes for a rigid body, let us consider it in a little more detail.

We know from (6.13) that $\dot{\mathbf{r}}_i^c$ must be the velocity of the ith particle relative to the centre of mass (see Figure 6.2). But, by (7.3), this velocity can be written in terms of the angular velocity $\boldsymbol{\omega}$ as

$$\dot{\mathbf{r}}_i^c = \boldsymbol{\omega} \times \mathbf{r}_i^c.$$

Hence,

$$T^c = \sum_{i=1}^{N} \frac{1}{2} m_i |\boldsymbol{\omega} \times \mathbf{r}_i^c|^2$$

$$= \sum_{i=1}^{N} \frac{1}{2} m_i s_i^2 \omega^2 \quad \text{from (7.4)}$$

$$= \frac{1}{2} I \omega^2 \quad \text{from (7.6)}.$$

So, the total kinetic energy of a rigid body combines a translational part and a rotational part and is given by

$$T = \frac{1}{2} M V^2 + \frac{1}{2} I \omega^2, \tag{7.24}$$

where $V = |\dot{\mathbf{R}}|$.

7.2.2 Instantaneous centre

Consider a rigid body (not necessarily a lamina) moving parallel to a fixed plane — the xy-plane, say. Let us fix axes $O'x'$ and $O'y'$ in the rigid body so that $O'x'y'$ is moving parallel to Oxy as depicted in Figure 7.12. For general motion of the rigid body (rotation and translation) there will, at any instant in time, be a point P, say, in $O'x'y'$ such that it is instantaneously at rest relative to Oxy. This point is called the *instantaneous centre*. The axis through the instantaneous centre and perpendicular to the plane is called the *instantaneous*

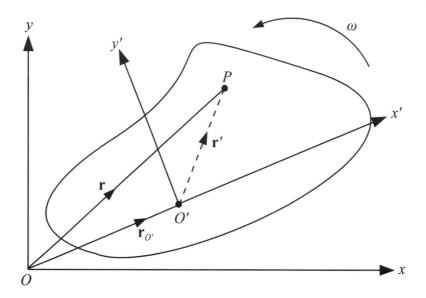

FIGURE 7.12
Instantaneous centre of a rigid body

axis about which rotation takes place. Although the instantaneous centre P
is pictured in the actual body in Figure 7.12, this may not necessarily be the
case for all time or indeed, any time. The location of the instantaneous centre
could be very far away from the body. Furthermore, if the motion of the rigid
body is purely translatory, $\omega = 0$, then the instantaneous centre is said to
reside at infinity.

As the rigid body moves along its path, the instantaneous centre also
traverses a path called the *space centrode* — the locus of P relative to Oxy.
The locus of P relative to $O'x'y'$ is called the *body centrode*. As the point of
contact between the space centrode and the body centrode is the instantaneous
centre, we say that *the body centrode rolls on the space centrode.*

To determine the position of the instantaneous centre relative to Oxy, we
refer to Figure 7.12.

Let $\mathbf{v}$ and $\mathbf{v}_{O'}$ be the velocities of P and O' relative to Oxy, respectively.
In general,

$$\mathbf{v} = \mathbf{v}_{O'} + \mathbf{v}',$$

where $\mathbf{v}' \equiv \dot{\mathbf{r}}'$ is the relative velocity of P with respect to O'. If it so happens
that the rigid body is rotating about an axis through O' then $\mathbf{v}' = \omega \times \mathbf{r}'$.
Thence,

$$\mathbf{v} = \mathbf{v}_{O'} + \omega \times \mathbf{r}'.$$

Now, let P be the instantaneous centre, implying that $\mathbf{v} = \mathbf{0}$; that is, at some instant in time P is that point about which the rigid body is rotating without translating. Therefore,

$$-\mathbf{v}_{O'} = \boldsymbol{\omega} \times \mathbf{r}' = \boldsymbol{\omega} \times (\mathbf{r} - \mathbf{r}_{O'}).$$

Taking the vector product with $\boldsymbol{\omega}$:

$$\boldsymbol{\omega} \times (-\mathbf{v}_{O'}) = \boldsymbol{\omega} \times [\boldsymbol{\omega} \times (\mathbf{r} - \mathbf{r}_{O'})]$$
$$= [\boldsymbol{\omega} \cdot (\mathbf{r} - \mathbf{r}_{O'})]\boldsymbol{\omega} - [\boldsymbol{\omega} \cdot \boldsymbol{\omega}](\mathbf{r} - \mathbf{r}_{O'}).$$

But $\boldsymbol{\omega}$ is perpendicular to the plane containing $\mathbf{r} - \mathbf{r}_{O'}$. Hence,

$$-\boldsymbol{\omega} \times \mathbf{v}_{O'} = -\omega^2(\mathbf{r} - \mathbf{r}_{O'})$$

or

$$\mathbf{r} = \mathbf{r}_{O'} + \frac{\hat{\mathbf{n}} \times \mathbf{v}_{O'}}{\omega}, \qquad (7.25)$$

where $\hat{\mathbf{n}}$ is the direction in which $\boldsymbol{\omega}$ points.

Example 7.6 A uniform disc of radius l rolls (without slipping) along a horizontal surface in the xy-plane (see Figure 7.13). Given that the angular velocity of the disc is $-\omega\mathbf{k}$ ($\mathbf{k}$ is a unit vector along the z-axis directed out of the page), locate the instantaneous centre, the space centrode and the body centrode.

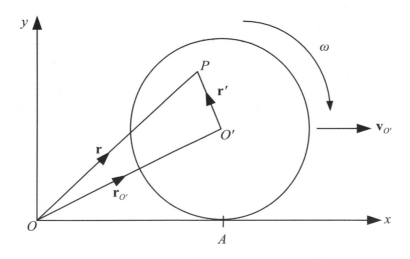

FIGURE 7.13
Rolling disc

Solution The motion of the disc is along the x-axis. This is because $\boldsymbol{\omega} = \omega(0, 0, -1)$ ensures a clockwise rotation. The disc rotates about an axis through the centre of mass at O'. The linear velocity of the disc is therefore $\mathbf{v}_{O'} = |\mathbf{v}_{O'}|(1,0,0) = (v_{O'},0,0)$. Substituting these values into (7.25) yields

$$\mathbf{r} = \mathbf{r}_{O'} + \frac{(0,0,-1) \times (v_{O'},0,0)}{\omega}$$

$$= (x_{O'},l,0) - (0,v_{O'}/\omega,0)$$

$$= (x_{O'},l - v_{O'}/\omega,0),$$

which is the position of the instantaneous centre.

As the disc moves, the instantaneous centre will, at all times, remain at a fixed horizontal distance $y = l - v_{O'}/\omega$ above the x-axis. The locus of this point — the space centrode — is, therefore, a line parallel to the x-axis. That the disc rolls without slipping implies that $v_{O'} = l\omega$ (this is the speed of O'),[2] which means that the space centrode is the x-axis itself. Hence, the instantaneous centre is located at the point of contact of the disc and the x-axis — A in Figure 7.13.

The body centrode is the locus of P relative to fixed axes in the disc. In this case it is given for $|\overrightarrow{O'P}| \equiv |\mathbf{r}'| = l$, which means that the locus of P is the circumference of the disc. $\quad\square$

7.2.3　The compound pendulum

A compound pendulum is a rigid body of mass M which under the action of a gravitational field g is free to oscillate in a plane about a fixed point O as depicted in Figure 7.14. Let the centre of mass be located at C, a distance l from O. The angle between OC and a vertical line passing through O is θ. Obtaining the equation of motion of the compound pendulum is facilitated using the principle of conservation of energy. The total kinetic energy[3] of the system is given by

$$T = \frac{1}{2}I\dot{\theta}^2, \tag{7.26}$$

where I is the moment of inertia of the compound pendulum about an axis through O and $\dot{\theta}$ is the angular speed. The potential energy of the system relative to the base point $V = 0$ is

$$V = -mgl\cos\theta. \tag{7.27}$$

Substituting (7.26) and (7.27) into the conservation of energy equation

$$T + V = E,$$

---

[2]The speed of the disc at different points will, of course, be different.

[3]We could have started with (7.24): $T = 1/2m(l\dot{\theta})^2 + 1/2I^c\dot{\theta}^2$, where $l\dot{\theta} = v$, the speed of the centre of mass, and I^c is the moment of inertia of the system about an axis through the centre of mass. Then employing the parallel axis theorem will again yield (7.26).

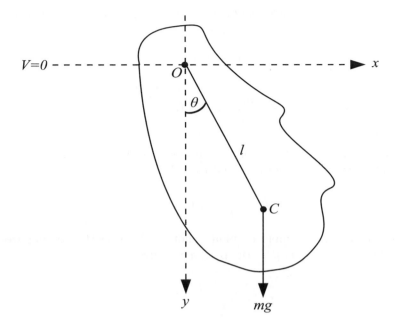

FIGURE 7.14
The compound pendulum

where E is the total constant energy, yields

$$\frac{1}{2}I\dot{\theta}^2 - mgl\cos\theta = E.$$

Differentiating this with respect to time gives

$$I\dot{\theta}\ddot{\theta} + mgl\dot{\theta}\sin\theta = 0$$

or

$$\ddot{\theta} + \frac{mgl}{I}\sin\theta = 0, \tag{7.28}$$

which is the equation of motion of the system.

The equation of motion (7.28) could equally have been arrived at using the torque equation (6.11):

$$\dot{\mathbf{L}} = \mathbf{G}.$$

Here, $\mathbf{L}$ is the angular momentum about an axis through O and $\mathbf{G}$ is the total external torque. Substituting (7.9) into this yields

$$I\ddot{\theta}\hat{\mathbf{n}} = \mathbf{r} \times \mathbf{F},$$

where $\ddot{\theta} \equiv \omega$, $\hat{\mathbf{n}}$ is the direction of the axis of rotation, which in this case is $(0, 0, -1)$, $\mathbf{r} \equiv \overrightarrow{OC} = (l \sin \theta, l \cos \theta, 0)$ and $\mathbf{F} = m\mathbf{g} = m(0, g, 0)$, which is taken in the direction of the y-axis. Thus,

$$I\ddot{\theta}(0, 0, -1) = (l \sin \theta, l \cos \theta, 0) \times (0, mg, 0)$$

implying

$$\ddot{\theta} + \frac{mgl}{I} \sin \theta = 0,$$

which is (7.28).

Of interest is the periodic time of small oscillations. For very small angular displacements $\sin \theta$ can be approximated to θ. Then

$$\ddot{\theta} + \frac{mgl}{I} \theta = 0,$$

which is the equation of simple harmonic motion. Writing the angular frequency of the swings as $\sqrt{mgl/I}$, the period τ is given by

$$\tau = 2\pi \sqrt{\frac{I}{mgl}}.$$

This is, in fact, the same period as a simple pendulum of length I/ml.

Example 7.7 A uniform solid sphere of mass m and radius l rolls without slipping down a rough inclined plane inclined at an angle α to the horizontal. Determine the equation of motion of the sphere.

Solution The principle of conservation of energy is the easiest method to employ (but see Exercise 7.8). The rolling sphere is depicted in Figure 7.15. Let x be the distance covered by the sphere in rolling from rest at O to A in time t. That the sphere rolls without slipping implies that A is a point of instantaneous rest, so $\dot{x} = l\omega$. Neither the normal reaction $\mathbf{N}$ nor the frictional force $\mathbf{f}$ (between the sphere and the plane) do work during the rolling process. This is because both these forces act through A, which is instantaneously at rest.

Using the conservation of energy equation,

$$T + V = E,$$

we have

$$T = \frac{1}{2} m\dot{x}^2 + \frac{1}{2} I\omega^2 = \frac{1}{2}\left(m + \frac{I}{l^2}\right)\dot{x}^2$$

and because the sphere is at a height $x \sin \alpha$ at A below O, the gravitational potential energy is

$$V = -mgx \sin \alpha.$$

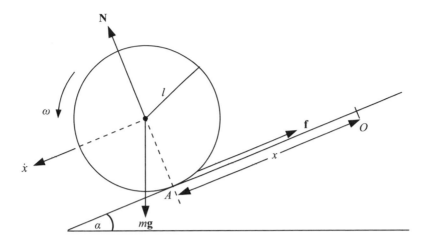

FIGURE 7.15
Sphere rolling down an incline

Hence

$$\frac{1}{2}\left(m + \frac{I}{l^2}\right)\dot{x}^2 - mgx\sin\alpha = E.$$

Differentiating this with respect to t yields

$$\left(m + \frac{I}{l^2}\right)\dot{x}\ddot{x} - mg\dot{x}\sin\alpha = 0.$$

Now, the moment of inertia of the sphere was found to be about $2ml^2/5$ an axis through its centre. Hence, on factoring out $\dot{x}$, the equation of motion can be written as

$$\ddot{x} - \frac{5}{7}g\sin\alpha = 0.$$

Note that had the sphere been hollow the moment of inertia would be $2ml^2/3$ (see Exercise 7.2). The equation of motion would then be

$$\ddot{x} - \frac{3}{5}g\sin\alpha = 0.$$

If for both cases, we applied the initial conditions $\dot{x}(0) = 0 = x(0)$ then the sphere with the greater moment of inertia (solid sphere) would have covered less distance than the sphere with the lesser moment of inertia (solid sphere) in equal time. $\square$

Example 7.8 A thin uniform plank of wood of mass m and length

and

$$N_y - mg = m\ddot{y} = \frac{mg}{4}\cos\theta(9\cos\theta - 6\cos\phi) - \frac{3mg}{4}$$

or

$$N_y = \frac{mg}{4}\cos\theta(9\cos\theta - 6\cos\phi) + \frac{mg}{4}.$$

Initially, $\theta(0) = \phi$, so the normal reactions at $t = 0$ are

$$N_x = \frac{3}{4}mg\sin\phi\cos\phi$$

$$N_y = mg\left(1 - \frac{3}{4}\sin^2\phi\right).$$

Now at some time $t > 0$, the top of the plank leaves the wall. This happens when $N_x = 0$. So, from (7.29), we must have the following condition satisfied:

$$\sin\theta(9\cos\theta - 6\cos\phi) = 0.$$

Clearly, $\sin\theta = 0$ implies that $\theta = 0$ and the plank lies vertically against the wall. Alternatively,

$$\cos\theta = \frac{2}{3}\cos\phi,$$

which corresponds to the plank having fallen by a third of its initial vertical height.

Notice that this result is independent of the length of the plank. Furthermore, if the initial angle $\phi = \pi/4$ say, the top of the plank would leave the wall when the angle between the plank and the wall was approximately 62°. □

7.3   Exercises

1.   Three particles P_1, P_2 and P_3 of mass m_1, m_2 and m_3, respectively, are joined by three light rods in such a way that P_1 P_2 P_3 are at the vertices of an equilateral triangle of side $2l$. Calculate the moment of inertia of this rigid body about (i) an axis through P_1 and perpendicular to the plane containing P_1 P_2 P_3, (ii) an axis passing through P_1 and the midpoint of the rod joining P_2 and P_3.

2.   Calculate the moment of inertia of a hollow sphere of mass m and radius l about a diameter.

3. A thin uniform rod has mass M and length $2l$. Calculate the moment of inertia of the rod about (i) an axis through the centre of the rod and at an angle $\alpha/2$ to the rod, (ii) an axis parallel to the rod and a distance d from the rod.

4. A uniform solid right circular cone has mass M, height h and radius l. Calculate the moment of inertia of the cone about (i) its axis of symmetry, (ii) an axis through its vertex and parallel to the base of the cone, (iii) an axis through a diameter of the cone's base.

5. A uniform block of mass M has edges of length $2a$, $2b$ and $2c$. A set of Cartesian axes $Oxyz$ is positioned inside the block such that the origin is at the centre of mass, and Ox, Oy and Oz are parallel to the edges of length $2a$, $2b$ and $2c$, respectively. Calculate the moment of inertia of the block about (i) the axis Oz, (ii) an axis along the edge of length $2b$.

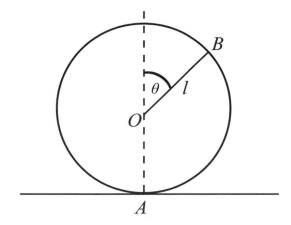

FIGURE 7.17
Disc rolling on a level surface

6. Consider the disc rolling on a surface in Figure 7.13 redrawn in Figure 7.17. Show that the point B, on the circumference of the disc, is travelling at speed $2l\dot{\theta}\cos(\theta/2)$. [Note that it will be useful to consider the point A in the disc; this is the point where the disc is in contact with the surface.].

7. Find the periodic time of small oscillations of a uniform rod of length $2l$ that is free to rotate about an axis perpendicular to the plane of rotation, which is located a distance d from the centre of mass.

8. Use the torque equation (6.19) to establish the equation of motion for the solid rolling sphere in Example 7.7.

8

Rotating Reference Frames

CONTENTS

Up until now, our discussion of the dynamical behaviour of particles and systems of particles has been with reference to some inertial frame; in this frame, Newton's second law applies. However, not all types of dynamical behaviour can be successfully modelled in inertial frames; indeed, it may be the case that employing a *non-inertial frame* of reference will yield more accurate quantitative results, as well as being more convenient. This is certainly true if one wishes to discuss particle dynamics close to the surface of the Earth. One would then choose a reference that is fixed relative to the Earth and, therefore, would rotate with the Earth. For example, in long range ballistics one must take account of the fact that the Earth is rotating and so adjustments would have to be made to any launching device to assure positional accuracy of a projectile. Moreover, readjustments would have to be made if the same projectile were to be launched from a different latitude.

In this chapter, we will be considering one particular type of non-inertial frame: the *rotating frame* of reference. For Newton's second law to remain valid in this frame, it will be reformulated to incorporate extra terms to account for the dynamical phenomena observed on a rotating Earth.

8.1 Rates of change in a rotating frame

Consider two Cartesian frames of reference S and S' in $\mathbb{E}^3$ as depicted in Figure 8.1. Let S be a fixed inertial frame (where Newton's second law holds) and S' be a non-inertial frame (where Newton's second law does not hold).

Assume that there exists a particle P that rotates about an axis through the

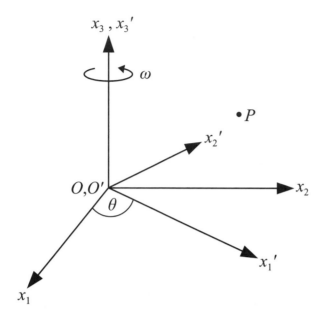

FIGURE 8.1
S' rotating relative to S about coincident axes

common origins O and O' in S and S', respectively. The position of P in S is given by

$$\mathbf{r} = x_1\mathbf{e}_1 + x_2\mathbf{e}_2 + x_3\mathbf{e}_3 \equiv x_i\mathbf{e}_i, \qquad i = 1, 2, 3. \qquad (8.1)$$

The position of P in S' is given by

$$\mathbf{r} = x_1'\mathbf{e}_1' + x_2'\mathbf{e}_2' + x_3'\mathbf{e}_3' \equiv x_i'\mathbf{e}_i', \qquad i = 1, 2, 3. \qquad (8.2)$$

Note that we are merely representing the same position vector $\mathbf{r}$ in two different coordinate systems.

Now, the vectors $\mathbf{e}_1$, $\mathbf{e}_2$ and $\mathbf{e}_3$ are unit basis vectors in the directions Ox_1, Ox_2 and Ox_3, respectively. They are, in effect, the standard Cartesian basis vectors $\mathbf{i}$, $\mathbf{j}$ and $\mathbf{k}$; however, they are written in such a way as to facilitate the use of the *Einstein summation convention*.

We will use this convention only in this chapter for reasons of economy and, hopefully, transparency. Simply put, but sufficient for our purpose, any term that possesses a repeated index, such as $x_i\mathbf{e}_i$, will automatically be assumed to sum over 1, 2 and 3. This obviates the requirement to incorporate $\sum$ in our notation. With this prescription, (8.1) follows. Everything said above applies equally well to primed quantities, and so (8.2) follows.

It was established in Section 7.1 that a particle with position vector $\mathbf{r}$ and free to rotate about an axis has a linear velocity $\dot{\mathbf{r}}$ that satisfies the equation (see (7.3))

$$\dot{\mathbf{r}} = \boldsymbol{\omega} \times \mathbf{r}, \tag{8.3}$$

where $\boldsymbol{\omega}$ is the angular velocity of the particle. This equation also holds for the motion of particle P depicted in Figure 8.1, where $\mathbf{r}$ and $\dot{\mathbf{r}}$ are measured with respect to O in S. Note that (8.3) holds for any vector fixed in a rotating system. In particular, it will hold for the unit basis vectors $\mathbf{e}'_i$ in S'. Thence

$$\dot{\mathbf{e}}'_i = \boldsymbol{\omega} \times \dot{\mathbf{e}}'_i. \tag{8.4}$$

What is the linear velocity of P, in Figure 8.1, in S? Well, the position of P in S is given by (8.1), but the basis vectors $\mathbf{e}_i$ are fixed for all time. This means that on differentiating (8.1) with respect to time the $\mathbf{e}_i$ remain constant. Thus

$$\dot{\mathbf{r}}_S = \dot{x}_i \mathbf{e}_i, \tag{8.5}$$

where $\dot{\mathbf{r}}_S$ denotes the velocity of P seen by someone sitting in the S frame observing P. What is the linear velocity of P in S'? The position of P in S' is given by (8.2) and the basis vectors $\mathbf{e}'_i$ are again fixed. Therefore, on differentiating (8.2) with respect to time, we find that

$$\dot{\mathbf{r}}_{S'} = \dot{x}'_i \mathbf{e}'_i, \tag{8.6}$$

where $\dot{\mathbf{r}}_{S'}$ denotes the velocity of P seen by someone sitting in the S' frame observing P.

Now, an observer fixed in S will see the basis vectors $\mathbf{e}'_i$ in S' change with time. So, differentiating $\mathbf{r}$ (as measured in S) — that is, differentiating (8.2) — yields

$$\dot{\mathbf{r}}_S = \dot{x}'_i \mathbf{e}'_i + x'_i \dot{\mathbf{e}}'_i.$$

With the aid of (8.4) and (8.6) this becomes

$$\dot{\mathbf{r}}_S = \dot{\mathbf{r}}_{S'} + x'_i (\boldsymbol{\omega} \times \mathbf{e}'_i)$$

or, with the aid of (8.2)

$$\dot{\mathbf{r}}_S = \dot{\mathbf{r}}_{S'} + \boldsymbol{\omega} \times \mathbf{r}. \tag{8.7}$$

The importance of this identity cannot be overstated. For, if $\mathbf{r}$ represented the position vector of a particle then (8.7) tells us how the velocity of the particle measured in the rotating frame can be interpreted in the fixed frame, and *vice versa*, where $\boldsymbol{\omega} \times \mathbf{r}$ is the velocity of the frame S' relative to the frame S. In this way it acts as a velocity transformation rule and holds even if $\boldsymbol{\omega}$ is a function of time. Furthermore, it also holds for any arbitrary vector, which means that it can be written as an operator:

$$\left(\frac{d}{dt} \right)_S = \left(\frac{d}{dt} \right)_{S'} + \boldsymbol{\omega} \times . \tag{8.8}$$

8.3   The centrifugal force

Physically, the centrifugal force

$$\mathbf{F}_{\text{centrifugal}} = -m\boldsymbol{\omega} \times (\boldsymbol{\omega} \times \mathbf{r}) \tag{8.13}$$

is familiar to anyone who has sat in a bus or car as it rounds a bend. The force that tends to push a passenger against the inside panel of a car or bus as it performs a manoeuvre is the centrifugal force. However, the panel exerts a reaction force that will, hopefully, keep the passenger in his seat. The acceleration associated with this reaction force is known as the *centripetal acceleration* $\boldsymbol{\omega} \times (\boldsymbol{\omega} \times \mathbf{r})$.

Consider the diagram in Figure 8.2. It could represent a small bob, P, attached to a piece of string itself attached to a fixed axis through O. As P

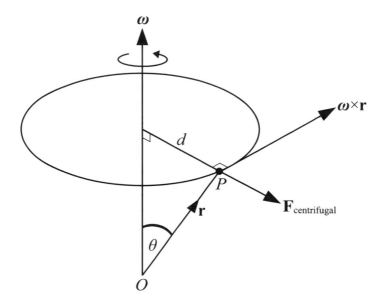

FIGURE 8.2
Centrifugal force

rotates about the axis the position, at all time, is $\mathbf{r}$. The angular velocity of P as it rotates in an anticlockwise direction is $\boldsymbol{\omega}$. The vector perpendicular to the plane containing $\boldsymbol{\omega}$ and $\mathbf{r}$ is $\boldsymbol{\omega} \times \mathbf{r}$, which is the linear velocity of P. If the string attached to P were suddenly severed the direction in which P would travel is $\boldsymbol{\omega} \times \mathbf{r}$. However, the force that tends to push P away from the axis of rotation is $\mathbf{F}_{\text{centrifugal}}$. As this force acts radially outwards it must be pointing in a direction both perpendicular to the axis of rotation and to $\boldsymbol{\omega} \times \mathbf{r}$. This direction is, therefore, $-\boldsymbol{\omega} \times (\boldsymbol{\omega} \times \mathbf{r})$.

The magnitude of the centrifugal force can be calculated easily. From (8.13)

$$\begin{aligned}
|\mathbf{F}_{\text{centrifugal}}| &= m|\boldsymbol{\omega} \times (\boldsymbol{\omega} \times \mathbf{r})| \\
&= m|\boldsymbol{\omega}||\boldsymbol{\omega} \times \mathbf{r}| \\
&= m|\boldsymbol{\omega}|^2|\mathbf{r}|\sin\theta \\
&= m\omega^2 d.
\end{aligned}$$

8.3.1 Actual and apparent gravitational acceleration

Consider a particle P falling freely under the influence of gravity close to the surface of a spherically symmetric Earth. What other forces may be acting? Certainly, there will be the centrifugal force due to the rotation of the Earth and the Coriolis force due to the particle's motion. Air resistance and perhaps other forces should be included. However, if it is assumed that the particle falls in a vacuum and that any 'other' forces may be neglected, there still remains the centrifugal force and Coriolis force. But, provided that the motion of the particle is considered just after release, the Coriolis force may also be neglected. So, for a particle of mass m, the equation of motion relative to an observer on the surface of the Earth is given by

$$m\ddot{\mathbf{r}} = \mathbf{F}_{\text{actual}} + \mathbf{F}_{\text{centrifugal}}, \tag{8.14}$$

where

$$\mathbf{F}_{\text{actual}} = m\mathbf{g}_{\text{actual}}.$$

We denote by $\mathbf{g}_{\text{actual}}$ the actual gravitational acceleration in the absence of the centrifugal force. The quantity $\mathbf{g}_{\text{actual}}$ is given by

$$\mathbf{g}_{\text{actual}} = -\frac{GM}{R^2}\hat{\mathbf{r}},$$

where M is the mass of the Earth, R is the radius of the Earth and $\hat{\mathbf{r}}$ is pointing radially outwards as depicted in Figure 8.3. Initially, P is subject to an apparent force $\mathbf{F}_{\text{apparent}}$ due to the presence of the centrifugal force $\mathbf{F}_{\text{centrifugal}}$. The right of (8.14) represents this apparent force, thus $\mathbf{F}_{\text{apparent}} = \mathbf{F}_{\text{centrifugal}} + \mathbf{F}_{\text{actual}}$. On dividing by m, we obtain the corresponding acceleration equation $\mathbf{g}_{\text{apparent}} = \mathbf{g}_{\text{actual}} - \boldsymbol{\omega} \times (\boldsymbol{\omega} \times \mathbf{r})$. The diagram in Figure 8.3 is, of course, exaggerated; we will find that the angle α between the actual and apparent gravitational accelerations will be extremely small.

Let us now balance the various accelerations in the directions of $\hat{\mathbf{r}}$ and $\hat{\boldsymbol{\theta}}$ so that the particle is in equilibrium. We have

$$g_{\text{apparent}} \cos\alpha = g_{\text{actual}} - \omega^2 R \cos^2\lambda \tag{8.15}$$

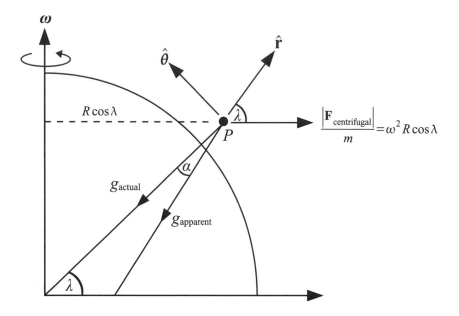

FIGURE 8.3
Particle P falling under the action of gravity close to the surface of the Earth

and
$$g_{\text{apparent}} \sin \alpha = \omega^2 R \cos \lambda \sin \lambda. \tag{8.16}$$
On dividing, we find that
$$\tan \alpha = \frac{\omega^2 R \cos \lambda \sin \lambda}{g_{\text{actual}} - \omega^2 R \cos^2 \lambda}. \tag{8.17}$$
Before we calculate α, a few things are required to be known about the right side of this equation.

The Earth rotates about its axis once every day relative to the Sun; this is a *solar day* comprising 24 hours or 86400 *mean solar seconds*.[4] Then the magnitude of the angular velocity of the Earth relative to an axis passing through the North and South Poles is
$$\omega = \frac{2\pi}{86400} \text{ rads}^{-1} \approx 7.3 \times 10^{-5} \text{ rads}^{-1}.$$
Taking the radius of the Earth R to be 6.4×10^6 m, we have
$$\omega^2 R \approx 3.4 \times 10^{-2} \text{ ms}^{-2}.$$

---

[4]A solar second is slightly longer than the conventional second in SI units. This is because the Earth's mean solar day has been gradually lengthening, and will continue to do so, due to the frictional forces associated with tides. Moreover, relative to the 'fixed' stars, the Earth rotates once every sidereal day, which is less than a mean solar day by approximately 3 minutes, 56 seconds.

Taking g_{actual} to be approximately 9.8 ms^{-2} (close to the Earth's surface), it is clear that $\omega^2 R \ll g_{\text{actual}}$, so (8.17) is approximately

$$\tan \alpha = \frac{\omega^2 R}{g_{\text{actual}}} \cos \lambda \sin \lambda = 3.5 \times 10^{-3} \cos \lambda \sin \lambda.$$

At the North Pole — that is at latitude $\lambda = \pi/2 - \alpha = 0$ and the actual and apparent gravitational accelerations coincide;

$$g_{\text{actual}} = g_{\text{apparent}}. \tag{8.18}$$

There is, of course, no centrifugal force at the North (South) Pole. The maximum deviation between the actual and apparent gravitational accelerations occurs when $\cos \lambda \sin \lambda$ is a maximum. This occurs when $\lambda = \pi/4$ giving an angle of deviation of

$$\alpha_{\text{max}} \approx 1.75 \times 10^{-3} \, \text{rad} \approx 0.1°.$$

On the equator, $\lambda = 0$, we have $\alpha = 0$. However, the centrifugal force would not, of course, vanish. So by (8.15)

$$g_{\text{apparent}} = g_{\text{actual}} - \omega^2 R. \tag{8.19}$$

From (8.18) and (8.19), we can compare the difference between the gravitational acceleration (actual or apparent) at the North (South) Pole and the equator. This value is $\omega^2 R = 3.4 \times 10^{-7}$ ms^{-2}. However, it has been experimentally determined that the difference is 5.2×10^{-2} ms^{-2} — around 53 percent larger than our calculated value. There is nothing alarming about this discrepancy, merely that the Earth is an oblate spheroid and so due to bulging at the equator (caused by the Earth spinning about its axis) the difference in radius at the pole and the equator will not be zero, as assumed above, but around 21 km.

Example 8.1 The cylindrical bucket depicted in Figure 8.4 is filled to a depth d with a fluid of mass m and density ρ. The bucket is set spinning about an axis of symmetry with angular velocity $\boldsymbol{\omega} = \omega \hat{\mathbf{z}}$. Use the equation of hydrostatics in the form

$$\frac{m}{\rho} \nabla p = \mathbf{F}_{\text{grav}} + \mathbf{F}_{\text{centrifugal}},$$

and the boundary condition $p(r, z) = p(0, d_{\text{min}}) = p_{\text{min}}$ to show that the pressure in the fluid is

$$p(r, z) = \frac{\rho \omega^2 r^2}{2} - \rho g(z - d_{\text{min}}) + p_{\text{min}}.$$

Then, use the boundary condition $p(r, Z(r)) = p_{\text{min}}$ to show that the shape of the free surface $Z(r)$ is parabolic. Finally, show that

$$d_{\text{min}} = d - \frac{\omega^2 l^2}{4g},$$

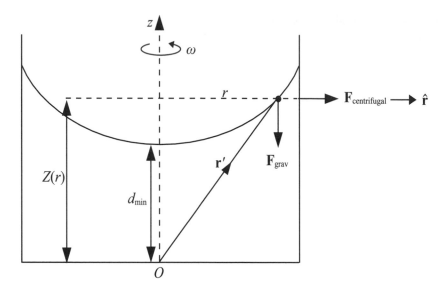

FIGURE 8.4
Rotating bucket

where l is the radius of the bucket.

Solution We are assuming that the bucket is fixed in a rotating frame S' relative to a fixed frame S. The fluid is also rotating at the same angular velocity as the bucket relative to S. Then the fluid is stationary relative to the bucket.

From the equation of hydrostatics, we have

$$\frac{1}{\rho}\boldsymbol{\nabla}p = \mathbf{g} - \boldsymbol{\omega} \times (\boldsymbol{\omega} \times \mathbf{r}')$$

$$= -g\hat{\mathbf{z}} - \omega^2\hat{\mathbf{z}} \times (\hat{\mathbf{z}} \times \mathbf{r}').$$

But $\hat{\mathbf{z}} \times (\hat{\mathbf{z}} \times \mathbf{r}') = (\hat{\mathbf{z}} \cdot \mathbf{r}')\hat{\mathbf{z}} - \mathbf{r}' = -\mathbf{r} = -r\hat{\mathbf{r}}$. Therefore,

$$\boldsymbol{\nabla}p = \rho(-g\hat{\mathbf{z}} + \omega^2 r\hat{\mathbf{r}}).$$

We now have three equations to solve. Writing $\boldsymbol{\nabla}$ in cylindrical polar coordinates:

$$\boldsymbol{\nabla} = \hat{\mathbf{r}}\frac{\partial}{\partial r} + \frac{\hat{\boldsymbol{\theta}}}{r}\frac{\partial}{\partial \theta} + \hat{\mathbf{z}}\frac{\partial}{\partial z},$$

and comparing coefficients of $\hat{\mathbf{r}}$, $\hat{\boldsymbol{\theta}}$ and $\hat{\mathbf{z}}$ yields

$$\frac{\partial p}{\partial r} = \rho\omega^2 r$$

$$\frac{1}{r}\frac{\partial p}{\partial \theta} = 0$$

$$\frac{\partial p}{\partial z} = -\rho g.$$

Integrating the second of these equations tells us that p is only a function of r and z. Integrating the first yields

$$p(r,z) = \frac{\rho\omega^2 r^2}{2} + A(z).$$

Differentiating this with respect to z and comparing with the third equation yields

$$\frac{dA}{dz} = -\rho g$$

or

$$A(z) = -\rho g z + B,$$

where B is an arbitrary constant of integration. Thus

$$p(r,z) = \frac{\rho\omega^2 r^2}{2} - \rho g z + B.$$

The boundary condition $p(0, d_{\min}) = p_{\min}$ gives $B = p_{\min} + \rho g d_{\min}$ resulting in the required pressure of the fluid

$$p(r,z) = \frac{\rho\omega^2 r^2}{2} - \rho g(z - d_{\min}) + p_{\min}.$$

The boundary condition $p(r, Z(r)) = p_{\min}$ gives the equation of the free surface

$$Z(r) = \frac{\omega^2 r^2}{2g} + d_{\min},$$

which is the equation of a parabola.

Finally, the value of $d_{\min}$ can be determined from the volume of fluid present in the bucket. So,

$$\text{volume} = \int_{\theta=0}^{2\pi}\int_{r=0}^{l} Z(r)r\,dr\,d\theta$$

$$= 2\pi \int_{r=0}^{l} \left(\frac{\omega^2 r^3}{2g} + d_{\min}r\right)dr$$

$$= \pi l^2 \left(\frac{\omega^2 l^2}{4g} + d_{\min}\right).$$

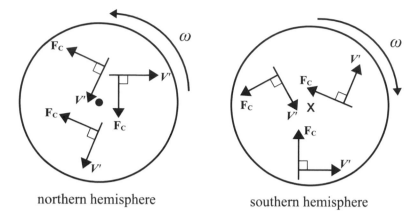

northern hemisphere   southern hemisphere

FIGURE 8.6
Two-dimensional description of the Coriolis force $\mathbf{F}_C$. The left disc is viewed
now directly above the North Pole and the right disc is viewed from directly
above the South Pole

bath or sink. However, the Coriolis force is not large enough to produce any
visible effect. So in practice there is just as much chance that the water will
drain in either a clockwise or anticlockwise direction.

8.4.1   Motion of a freely falling body

Objects dropped from a high tower — the Leaning Tower of Pisa, for example
— may not hit the ground at precisely the spot that your intuition would have
you believe. Inevitably, one must take into consideration both the centrifugal
and Coriolis forces to determine the final position of a falling object with
a 'reasonable' degree of accuracy. To make our analysis a little easier, we
will neglect air resistance. It will be further assumed that the Earth's axis
of symmetry does not process or *nutate*,[6] which means that the Euler force
vanishes: $\dot{\boldsymbol{\omega}} = \mathbf{0}$.

The equation of motion, in a rotating frame, for a body of mass m falling
to Earth is from (8.12)

$$\mathbf{F}' = \mathbf{F} + \mathbf{F}_{\text{Coriolis}} + \mathbf{F}_{\text{centrifugal}},$$

or, on dividing through by m,

$$\ddot{\mathbf{r}}' = \mathbf{g} - 2\boldsymbol{\omega} \times \dot{\mathbf{r}}' - \boldsymbol{\omega} \times (\boldsymbol{\omega} \times \mathbf{r}).$$

---

[6]Nutation occurs when the Earth's axis oscillates due to the plane of the Moon's orbit
around the Earth being tilted 5.1° relative to the plane of the Earth's orbit around the Sun.
The British astronomer James Bradley (1693-1762) discovered this phenomenon in 1748.

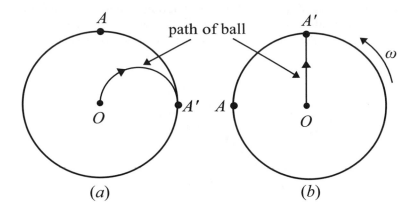

(a)   (b)

FIGURE 8.7
Rotating bowling alleys as seen by an observer in (a) a frame rotating with
the alley and (b) a frame fixed to the bowling ball

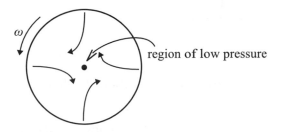

region of low pressure

FIGURE 8.8
Air particles being deflected to the right surrounding a region of low pressure
and creating a vortex that rotates in an anticlockwise direction

This equation can be reduced further if our high tower is not stratospher-
ically high (the Leaning Tower of Pisa is about 55 m in height). So as the
ratio of centrifugal force to gravitational force is negligible — as was shown
in Section 8.31 — we are at liberty to ignore the $-\boldsymbol{\omega} \times (\boldsymbol{\omega} \times \mathbf{r})$ term resulting
in the much simpler equation

$$\ddot{\mathbf{r}}' = \mathbf{g} - 2\boldsymbol{\omega} \times \dot{\mathbf{r}}'.$$

Because there is no possibility of confusion arising, we will drop the primes
from the equation above and consider instead

$$\ddot{\mathbf{r}} = \mathbf{g} - 2\boldsymbol{\omega} \times \dot{\mathbf{r}}. \tag{8.22}$$

However, do not forget that this is still the equation of motion in a rotating
frame S'.

To aid our analysis of a falling body, a diagram representing the various compass points and salient vector directions will prove useful. This is depicted in Figure 8.9. Axes $Qxyz$ are fixed to the surface of the rotating Earth such

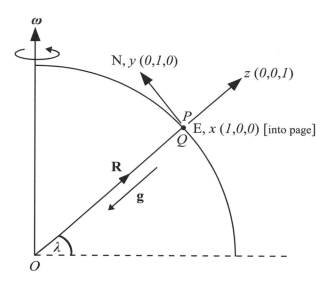

FIGURE 8.9
Body located at $p(x, y, z)$ above the surface of the Earth and falling towards the Earth

that Qy is pointing north (N), Qz is pointing radially outwards and Qx is pointing east (E) into the page. The body, initially at $p(x, y, z)$, is at latitude λ in the northern hemisphere, and is falling radially inwards in the direction $-\mathbf{R}$. The unit basis vectors in the directions Qx, Qy and Qz are, respectively, $(1, 0, 0)$, $(0, 1, 0)$ and $(0, 0, 1)$. The direction of the free-fall acceleration $\mathbf{g}$ is parallel to $-\mathbf{R}$ and given by $\mathbf{g} = (0, 0, -g)$. The position vector $\mathbf{R}$ is fixed with constant magnitude R representing the radius of a spherical Earth.

Now, the angular velocity $\boldsymbol{\omega}$ acts in the Qyz-plane and has no component in the x-direction. Thus resolving in these directions, we have

$$\boldsymbol{\omega} = (0, \omega \cos \lambda, \omega \sin \lambda). \tag{8.23}$$

The position vector $\overrightarrow{QP} \equiv \mathbf{r} = (x, y, z)$ is not shown in Figure 8.9. On substituting all of the relevant information into (8.22) and comparing components, we obtain the following three second-order differential equations:

$$\ddot{x} = 2\omega(\dot{y}\sin\lambda - \dot{z}\cos\lambda) \tag{8.24}$$

$$\ddot{y} = -2\omega\dot{x}\sin\lambda \tag{8.25}$$

$$\ddot{z} = -g + 2\omega\dot{x}\cos\lambda. \tag{8.26}$$

Assume that the body is released at a point $(0, 0, h)$ relative to the origin Q, at time $t = 0$. On integrating (8.24) and (8.25), we find that

$$\dot{x} = 2\omega(y \sin \lambda - z \cos \lambda) + C_x$$

and

$$\dot{y} = -2\omega x \sin \lambda + C_y,$$

where C_x and C_y are constants of integration. From the initial conditions $x = 0 = y$, $\dot{x} = 0 = \dot{y}$ and $z = h$, we find that

$$C_x = 2\omega h \cos \lambda \quad \text{and} \quad C_y = 0.$$

Substituting back gives

$$\dot{x} = 2\omega(y \sin \lambda + (h - z) \cos \lambda)$$

and

$$\dot{y} = -2\omega x \sin \lambda; \tag{8.27}$$

therefore

$$\ddot{z} = -g + (2\omega)^2 (y \sin \lambda + (h - z) \cos \lambda) \cos \lambda.$$

But $O|\omega|^2 \ll g$, which implies that

$$\ddot{z} = -g$$

or

$$\dot{z} = -gt + C_z,$$

where C_z is a constant of integration. Assuming $\dot{z} = 0$ at $t = 0$, we have $C_z = 0$. So

$$\dot{z} = -gt \tag{8.28}$$

or

$$z = -\frac{1}{2}gt^2 + h \tag{8.29}$$

on evaluating the constant of integration at $t = 0$.

Substituting (8.27) and (8.28) into (8.24) yields

$$\ddot{x} = 2\omega gt \cos \lambda,$$

where terms $O|\omega|^2$ have been neglected. Integrating this equation twice gives

$$x = \frac{1}{3}\omega gt^3 \cos \lambda. \tag{8.30}$$

Now, from (8.29), the time taken for a body to fall through a height h and hit the ground ($z = 0$) is

$$t = \sqrt{\frac{2h}{g}}. \tag{8.31}$$

Substituting (8.31) into (8.30) yields

$$x = \frac{1}{3}\omega g \left(\frac{2h}{g}\right)^{3/2} \cos \lambda. \tag{8.32}$$

Thus, it can be readily seen that when the body hits the ground it will have been displaced slightly in the x-direction; that is, east. So for a body dropped from the Leaning Tower of Pisa (Pisa is around latitude 44°) a height of 55 m above the ground, it will be deflected to the east a distance

$$d = \frac{1}{3}(7.3 \times 10^{-5})(9.8)(11.2)^{3/2} \cos 44°$$

$$\approx 6.5 \, \text{mm}.$$

This effect can also be understood in the inertial frame fixed to and rotating with the Earth. As the Earth rotates in an easterly direction the body, just as it begins to fall, also rotates east. By conservation of angular momentum the body must increase in speed, but in the east direction.

8.4.2   Foucault's pendulum

On the 31 March 1851, beneath the great Panthéon dome, an enormous pendulum was set in motion. The dense bob of the pendulum, consisting of a lead ball encased in brass, weighed 28 kg, had a radius of 19 cm and was suspended from a steel wire that measured 67 m. "After a double oscillation lasting sixteen seconds, we saw it return approximately 2.5 millimetres to the left of its starting point. As the same effect continued to occur with each oscillation, the deviation was ever increasing, in proportion to the passing of time." These were the words spoken by the French physicist Jean Bernard Léon Foucault (1819-1868) as he witnessed the magnificent device that now bears his name: the *Foucault pendulum*.

   The peculiar motion that Foucault describes is the first observational evidence of the effect of the Coriolis force and demonstrates that the Earth does indeed rotate about its own axis.

   Let us first derive the equations of motion for Foucault's pendulum and then solve them for appropriate initial conditions.

   The arrangement of the pendulum is such that it is free to swing in any vertical plane. So that the motion may continue for a protracted period of time, the supports are designed to reduce as much of the frictional resistance as possible. The Foucault pendulum as depicted in Figure 8.10 has its origin O positioned in its equilibrium position; that is, the bob B resides at O when the pendulum is motionless. The pendulum is free to swing in any plane perpendicular to the xy-plane, where the x-axis is pointing east, the y-axis is pointing north and z points radially out (up). The tension in the thin wire of length l is denoted as $\mathbf{f}_T$ and is directed to the point of suspension at A. Thus $OA = OB = l$. The standard basis vectors $\mathbf{e}_1$, $\mathbf{e}_2$ and $\mathbf{e}_3$ describe a right-handed coordinate system.

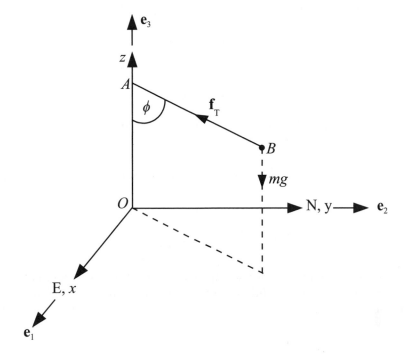

FIGURE 8.10
Foucault's pendulum free to swing in any plane perpendicular to xy-plane

Assume that the motion is initially simple harmonic (ϕ small). Projecting $\mathbf{f}_T$ in the $\mathbf{e}_1$, $\mathbf{e}_2$ and $\mathbf{e}_3$ directions yields

$$\begin{aligned}
\mathbf{f}_T &= (\mathbf{f}_T \cdot \mathbf{e}_1)\mathbf{e}_1 + (\mathbf{f}_T \cdot \mathbf{e}_2)\mathbf{e}_2 + (\mathbf{f}_T \cdot \mathbf{e}_3)\mathbf{e}_3 \\
&= (\mathbf{f}_T \cos\theta_1)\mathbf{e}_1 + (\mathbf{f}_T \cos\theta_2)\mathbf{e}_2 + (\mathbf{f}_T \cos\theta_3)\mathbf{e}_3,
\end{aligned}$$

where θ_i ($i = 1, 2, 3$) are the angles between $\mathbf{f}_T$ and $\mathbf{e}_i$. Hence,

$$\mathbf{f}_T = -\frac{\mathbf{f}_T x}{l}\mathbf{e}_1 - \frac{\mathbf{f}_T y}{l}\mathbf{e}_2 + \frac{\mathbf{f}_T(l-z)}{l}\mathbf{e}_3.$$

Because the centrifugal and Euler forces are negligible here, the equation of motion for this system is by (8.12)

$$\mathbf{F}' = \mathbf{F} + \mathbf{F}_{\text{Coriolis}}$$

or

$$m\ddot{\mathbf{r}}' = (\mathbf{f}_T + m\mathbf{g}) - 2m\boldsymbol{\omega} \times \mathbf{v}'$$

or on dropping the primes (but remember we are discussing motion with respect to a rotating frame S')

$$m\dot{\mathbf{r}} = (\mathbf{f}_T + m\mathbf{g}) - 2m\boldsymbol{\omega} \times \mathbf{v}. \tag{8.33}$$

For the pendulum located at a latitude λ, we have from (8.23)

$$\boldsymbol{\omega} = (0, \omega \cos \lambda, \omega \sin \lambda).$$

Also, as gravity is $\mathbf{g} = -g\mathbf{e}_3$, we then have the three components of the equation of motion as follows:

$$m\ddot{x} = -\frac{f_T x}{l} - 2m\omega(\dot{z} \cos \lambda - \dot{y} \sin \lambda) \tag{8.34}$$

$$m\ddot{y} = -\frac{f_T y}{l} - 2m\omega \dot{x} \sin \lambda \tag{8.35}$$

$$m\ddot{z} = \frac{f_T(l - z)}{l} - mg + 2m\omega \dot{x} \cos \lambda. \tag{8.36}$$

It is now necessary to make a few approximations. If ϕ is considered to be very small then all of the pendulum motion can be assumed to be confined to the xy-plane. This means that z, $\dot{z}$ and $\ddot{z}$ all vanish. Equation (8.36) then becomes

$$f_T = mg - 2m\omega \dot{x} \cos \lambda,$$

and on substituting this into (8.34) and (8.35), we have

$$\ddot{x} = -\frac{gx}{l} + \frac{2\omega x \dot{x} \cos \lambda}{l} + 2\omega \dot{y} \sin \lambda$$

$$\ddot{y} = -\frac{gy}{l} + \frac{2\omega y \dot{x} \cos \lambda}{l} - 2\omega \dot{x} \sin \lambda.$$

These two equations are slightly unwieldy due to the presence of the non-linear terms $x\dot{x}$ and $y\dot{x}$. However, as we are considering x, y and ω to be small, the two non-linear terms can be considered negligible by comparison. Thence,

$$\ddot{x} = -\frac{gx}{l} + 2\omega \dot{y} \sin \lambda$$
$$\ddot{y} = -\frac{gy}{l} - 2\omega \dot{x} \sin \lambda. \tag{8.37}$$

By employing appropriate initial conditions, we can solve these coupled linear second-order differential equations.

Now, to solve the system (8.37), we will employ the following technique. Define a new complex variable $\zeta(t)$ such that

$$\zeta = x + iy \tag{8.38}$$

and

$$\dot{\zeta} = \dot{x} + i\dot{y}. \tag{8.39}$$

Differentiating this twice with respect to t and substituting in (8.37) as appropriate yields

$$\ddot{\zeta} = -\frac{g}{l}(x + iy) + 2\omega \sin \lambda (\dot{y} - i\dot{x})$$

$$= -\frac{g}{l}\zeta - 2i\omega \sin \lambda \dot{\zeta}$$

or

$$\ddot{\zeta} + 2i\omega \sin \lambda \dot{\zeta} + \frac{g}{l}\zeta = 0. \tag{8.40}$$

Next, assume a solution of the form $\zeta = e^{kt}$ for some constant k.

On substituting this in, the following characteristic equation is obtained:

$$k^2 + 2i\omega \sin \lambda k + \frac{g}{l} = 0.$$

(Note that the fact that this equation contains complex constants in no way interferes with the solution process, as they are treated like any other constants.)

On solving this quadratic equation, we obtain

$$k = -i\omega \sin \lambda \pm i\sqrt{\omega^2 \sin^2 \lambda + g/l}$$

$$\approx -i\omega \sin \lambda \pm i\sqrt{g/l}.$$

Recall that in this scenario $\omega^2 \approx 0$. The general solution to (8.40) is then

$$\zeta = e^{i\omega t \sin \lambda}\left(C_1 e^{it\sqrt{g/l}} + C_2 e^{-it\sqrt{g/l}}\right), \tag{8.41}$$

where C_1 and C_2 are arbitrary constants. To determine the constants C_1 and C_2, we will assume that at $t = 0$ the pendulum is released from rest at position $(x, y) = (A, 0)$ — that is, it is displaced a distance A to the east before being released. These initial conditions give $C_1 = C_2 = A/2$. To verify that this is the case, compare (8.38) with (8.41) and (8.39) with the derivative of (8.41) once the initial conditions have been imposed. The solution (8.41) can now be written as

$$\zeta = Ae^{i\omega t \sin \lambda} \cos\left(\sqrt{\frac{g}{l}}t\right).$$

At $t \approx 0$, the phase factor $e^{-i\omega t \sin \lambda}$ is approximately unity and the pendulum oscillates with simple harmonic motion, the amplitude of the motion being A and the angular frequency $\sqrt{g/l}$:

$$x = A\cos\left(\sqrt{\frac{g}{l}}t\right).$$

Thus the pendulum oscillates east-west. However, as time increases the phase factor becomes more prominent and the plane of motion begins to rotate through a phase angle of $\omega t \sin \lambda$ in the xy-plane. This corresponds to a clockwise rotation (viewed from above) in the northern hemisphere, and an anticlockwise rotation in the southern hemisphere.

If the pendulum were located at the North Pole, say, $\lambda = \pi/2$ giving a phase factor of $e^{-i\omega t}$. This means that the pendulum will rotate at the same rate as the Earth. At the equator, $\lambda = 0$, there is no phase factor and the pendulum will oscillate without rotation. The latitude of Paris is approximately $49°$, so the period of Foucault's pendulum inside the Panthéon is

$$\tau = \frac{2\pi}{\omega \sin \lambda} \approx \frac{2\pi}{(7.3 \times 10^{-5})(0.75)} \approx 114045 \text{ seconds,}$$

which is approximately 32 hours.

8.5   Exercises

1. Consider two Cartesian frames S (fixed) and S' (rotating with respect to S) that have the same origin. Relative to an observer fixed in S', a particle P has position $\mathbf{r} = (2t, -t, t^2)$, where t represents time. If the angular velocity of S' relative to S is $\boldsymbol{\omega} = (t, -t^2, 3t+1)$, find the velocity of P as seen by an observer in S', and determine how this velocity is interpreted in S when $t = 1$.

2. With reference to the frames S and S' in Exercise 8.1, a particle has position $\mathbf{r} = (\sin t, -\cos t, t)$. If the angular velocity of S' relative to S is $\boldsymbol{\omega} = (\cos t, \sin t, 1)$, find the actual and apparent accelerations of P for $t > 0$.

3. Consider two Cartesian frames S and S' that do not have the same origin. Instead, the position of the origin O' in S' is $\mathbf{R}$ relative to O in S. If a particle P has position $\mathbf{r}$ in S and $\mathbf{r}'$ in S', show that the acceleration of P as observed in S is

$$\ddot{\mathbf{r}} = \ddot{\mathbf{R}} + \ddot{\mathbf{r}}' + \dot{\boldsymbol{\omega}} \times \mathbf{r}' + 2\boldsymbol{\omega} \times \dot{\mathbf{r}}' + \boldsymbol{\omega} \times (\boldsymbol{\omega} \times \mathbf{r}'),$$

where $\ddot{\mathbf{R}}$ is the acceleration of O' with respect to O.

4. A ship has mass 10^4 kg. Find (correct to two significant figures) the magnitude of the centrifugal force acting on the ship at latitude $\lambda = 0$ and $\lambda = \pi/3$.

5. A river located at latitude λ flows in a northerly direction. If the speed of the flowing water is v and the distance between the east and west banks of the river is d, show that the difference between the water level on the east bank and that of the west bank is

$$\frac{2d\omega v \sin \lambda}{\sqrt{g^2 + 4\omega^2 v^2 \sin^2 \lambda}},$$

where g is the gravitational acceleration and ω is the Earth's angular velocity.

6. A smooth straight wire Ox of length $2l/\sqrt{3}$ rotates in the xy-plane with constant angular speed ω about the z-axis through O. A small bead is threaded onto the wire so that it is initially at rest a distance $l/\sqrt{3}$ from O. Show that

$$\ddot{x} - \omega^2 x = 0.$$

How long will it take for the bead to reach the end of the wire? At what speed, relative to the wire, will the bead leave the wire?

7. A particle is projected vertically (radially outwards) from a point on the Earth's surface. If the speed of projection is v and the point of projection is at latitude λ, show that it will impact the Earth west and a distance

$$\frac{4}{3g^2} \omega v^3 \cos \lambda$$

from its initial projection.

8. A particle is projected from the Earth's surface at an angle of elevation θ. The point of projection is at latitude λ and it initially travels due south with speed v. Show that the particle will impact the Earth west and at a distance

$$\frac{4}{3g^2} \omega v^3 \sin^2 \theta (\sin \theta \cos \lambda + 3 \cos \theta \sin \lambda)$$

from its initial projection.

9. The particle in Exercise 8.8 is now projected due east. If all other conditions remain the same as in the previous exercise, in what direction will the Earth's rotation cause the particle to deviate and by what distance?

Part II

Relativity

9

Special Relativity

CONTENTS

9.1 Inception

The period following the publication of Newton's *Principia* was an extremely fruitful one for classical dynamics. A great many advances in its formulations were developed by such notables as Joseph Louis Lagrange (1736-1813) and William Rowan Hamilton (1805-1865), notwithstanding the numerous predictions of an astronomical nature and otherwise. In short, Newtonian theory was considered to be invincible.

The concept of a vector field was certainly not a new one for the Victorian physicist or mathematician. Indeed, they were very familiar with magnetic, electric and gravitational fields surrounding magnets, charged bodies and massive bodies, respectively. However, the fields by themselves were not endowed with any physical structure — they were merely an arena where forces could act upon particles. But these ideas were about to change through the pioneering work of two people in particular: the English experimentalist Michael Faraday (1791-1867) and the Scottish theoretician James Clerk Maxwell (1831-1879).

The mathematical equations devised by Maxwell with the aid of the work of Faraday — known as *Maxwell's equations* — suggested a possible conflict with the established Newtonian structure. Rather than the electromagnetic field act as some kind of arena, the electric and magnetic fields were found to interact with each other. For example, one of the oscillating fields would give rise to the other oscillating field and *vice versa*. Moreover, the speed of propagation of these effects (*in vacuo*) was found to be the *speed of light*: $c = 3 \times 10^8$ ms^{-1}.

One startling consequence of Maxwell's equations was their dogged refusal to be invariant under the Galilean transformations. This was, in no uncertain terms, extremely vexing for the Victorian scientific community. Aside from the fact that the so-called null experiment of 1887 performed by Albert Abraham Michelson (1852-1931) and Edward Williams Morley (1838-1923) to measure the Earth's motion through the aether(the medium through which light rays and, therefore, electromagnetic waves were assumed to propagate) only exacerbated matters by failing to detect any such motion.

9.1.1   The Michelson–Morley experiment (a précis)

The effect of disturbances in the electromagnetic field propagate at the speed of light. Maxwell concluded that the medium of propagation was the so-called *luminiforous aether* and that the aetherial frame was that of the 'fixed stars'. This is the absolute space of Newtonian dynamics.

Undoubtedly, the most celebrated of all the aether experiments is the Michelson–Morley experiment.[1]

The apparatus that they adopted to detect the so-called 'aether wind'[2] is called an *interferometer* — invented by Michelson some years before. Essentially, a source of light in the form of a narrow beam was split into two parts. One beam passed through a semi-transparent plate, inclined at an angle of 45° to the beam, and was reflected by a mirror so that it traversed the same path, but in the opposite direction, and then struck the same semi-transparent plate again. Part of this beam was reflected by the plate and passed through a telescope. The beam that arrived at the telescope was perpendicular to the beam that was reflected by the mirror. The second beam was reflected by the plate at a 90° angle and was then reflected by another mirror. The beam traversed the same path before passing through the semi-transparent plate and on to the telescope. A schematic drawing is given in Figure 9.1. When viewed through the telescope, interference fringes would be observed provided that the mirrors were not absolutely orthogonal to each other. The apparatus was then rotated through an angle of 90° so that the beam of light would travel at

---

[1]This experiment was allegedly proposed by Maxwell, but the technology of the period was not sophisticated enough for the experiment to be performed. I would like to thank Professor Hall for bringing this to my attention.

[2]The aether wind 'blows' past the apparatus as the Earth passes through the aether.

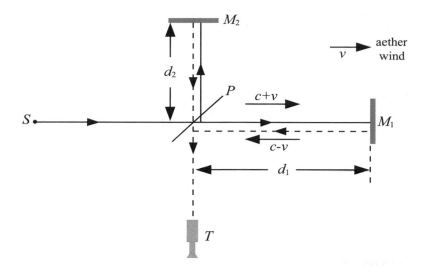

FIGURE 9.1
Schematic of Michelson's interferometer. M_1 and M_2 represent mirrors, S is the source of light, T is the telescope and P is the semi-transparent plate

right angles to the aether wind, and the experiment was then repeated. What was sought (but never found) was a slight fringe shift.[3]

Now, let d_1 be the distance between P and M_1, and d_2 be the distance between P and M_2. Assuming that the speed of light relative to the aether is c, then the times taken for the beam to go from P and reflect off the mirrors and return to P, are

$$t_1 = \frac{2d_1/c}{1 - v^2/c^2}, \quad t_2 = \frac{2d_2/c}{(1 - v^2/c^2)^{1/2}},$$

where v is the velocity of the aether wind. Provided $d_1 = d_2$ a value of v can be obtained in principle from the ratio

$$\frac{t_1}{t_2} = \sqrt{\frac{1}{1 - v^2/c^2}}.$$

(In practice, the difference is taken, which leads to a shift in the interference pattern.) The experiment amounts to the fact that $t_1 = t_2$ implying that v is very nearly zero, whereas it was assumed to be around 3×10^4 ms^{-1} (the orbital velocity of the Earth around the Sun).

Although for Michelson and Morley this was a very disappointing result, it

---

[3]The expected value was 0.4 of a fringe; instead the shift was less than 5×10^{-3} of a fringe, which was well within experimental error.

did herald a number of attempts to reconcile the findings with the prevailing dynamical theory.

A rather ingenious way of interpreting the null result was published in 1889 by the Irish theoretical physicist George Francis Fitzgerald (1851-1901). His explanation was that when a body moves through the aether with speed v, the length of the body in the direction of motion diminishes by an amount $\sqrt{1 - v^2/c^2}$. This means that l_1, in Figure 9.1, will be shorter than l_2 such that

$$l_1 = l_2\sqrt{1 - v^2/c^2}. \tag{9.1}$$

In 1904 the Dutch physicist Hendrik Anton Lorentz (1853-1928) arrived at the same result and developed the idea into what is called his *aether theory*. Equation (9.1) is now referred to as the *Lorentz–Fitzgerald contraction*.

9.2 Einstein's postulates of special relativity

Remarkably, Einstein claimed to be completely unaware of the Michelson–Morley experiment and its findings, insisting instead that his path towards his own relativistic theory had been laid down by the invariance of Maxwell's equations under Galilean transformations. Einstein was by no means alone in pondering over this perplexing anomaly. Another giant of the scientific community, the French mathematician Jules Henri Poincaré (1854-1912), had also given serious consideration to this matter. Indeed, they both independently discovered a relativity principle that Maxwell's equations satisfied.

Special relativity[4] relies upon two fundamental postulates; in fact, as we will shortly see, the first postulate implies the second.

- *Postulate* 1: (The principle of relativity), all inertial frames[5] are equivalent with respect to all the laws of physics.

- *Postulate* 2: The speed of light *in vacuo* has the same value in all inertial frames.

If one looks a little closer at Postulate 1 it will be seen to be quite a daring statement to propound. For it implies that *all* physics — including future physics — *must* adhere to the postulate. As such, the unity of physics must prevail — no physical theory can be independent of any other. So far, this postulate has remained intact, even though many ardent challenges have tried to compromise its integrity.

It is only when one considers Postulate 2 that all common sense breaks

---

[4]The theory is called 'special' because it does not contain a mechanism for gravitation. 'General' relativity incorporates the gravitational field.

[5]These frames are inertial in the Newtonian sense.

down, and clashes violently with our understanding of Newtonian dynamics. If I light a candle, the emitted light will recede from the source at speed c. If I were to run so as to catch up with the receding light it would still be receding from me at speed c. Even if I were somehow able to transport myself to an inertial frame travelling at near the speed of light and gave chase, the light would still be receding from me at speed c! There is no commonsense explanation for this effect, but this is precisely implied by the Michelson–Morley experiment.

9.3 Lorentz transformations

Although Einstein is credited with being responsible for the first derivation of the Lorentz transformations from the relativity principle, he was not the first to postulate the existence or form of these transformations. Indeed, as early as 1895, Poincaré predicted the non-existence of the aether frame and a number of years later, but before Einstein, utilised the *Lorentz transformations* to illustrate that a new dynamical theory was required that contained an upper velocity limit, c. However, it was Lorentz in 1903 (two years before the publication of Einstein's five ground-breaking papers, two of which had nothing to do with relativity) who formulated the equations that now bear his name.

Consider two inertial frames S and S' in *standard configuration* as depicted in Figure 9.2. This is called standard configuration if S and S' move relative to each other with speed v, the axes x and x' are coaxial and the origins of S and S' coincide at $t = 0 = t'$. As is evident from the above discussion the Galilean

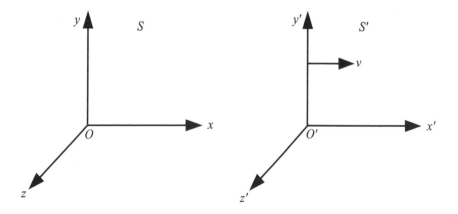

FIGURE 9.2
Inertial frames in standard configuration

transformations (1.4) no longer hold when speeds comparable with the speed of light c are considered. To derive an appropriate set of transformations that will fulfil all of the necessary criteria as mentioned in Section 9.2, we will assume a general set of transformations of the form

$$
\begin{aligned}
x' &= f_1(x, y, z, t) \\
y' &= f_2(x, y, z, t) \\
z' &= f_3(x, y, z, t) \\
t' &= f_4(x, y, z, t)
\end{aligned}
\tag{9.2}
$$

for some functions f_1, f_2, f_3 and f_4.

It can be deduced that the transformations are linear if we assume that a particle in S travelling at constant velocity $\mathbf{u}$ will in S' be travelling at constant velocity $\mathbf{u}'$; if this were not the case then Postulate 1 would not hold. Furthermore, since the inertial frames in Figure 9.2 are in standard configuration the y' and z' dependency in (9.2) can be greatly reduced. For if an event occurs in the $y = 0$ plane it must also occur in the $y' = 0$ plane. Thus, bearing in mind linearity, the y' equation can be written

$$
0 = a_1 x + a_3 z + a_4 t
$$

for constants a_1, a_3 and a_4 that may depend on v.

This equation must be satisfied identically for all x, z and t, implying that a_1, a_3 and a_4 are identically zero. Thus, we can now write the y' equation as

$$
y' = a_2 y
$$

for constant a_2. Reversing the directions of all the axes in Figure 9.2 except the y and y' axes yields, by symmetry and isotropy of space,

$$
y = a_2 y'.
$$

Thence, $a_2 = \pm 1$. The origins of S and S' will coincide at $t = 0 = t'$. However, if $v = 0$ we must have

$$
y' = y,
\tag{9.3}
$$

so $a_2 = 1$, and by an equivalent argument

$$
z' = z.
\tag{9.4}
$$

An event at $x' = 0$ in S' will correspond to $x = vt$ in S. Thus, the x' equation in (9.2) can be written as

$$
0 = b_1 v t + b_2 y + b_3 z + b_4 t,
$$

for constants b_1, b_2, b_3 and b_4 that may depend on v. This must be identically satisfied for all y, z and t. Hence $b_2 = 0 = b_3$ and $b_4 = -b_1 v$, yielding

$$
x' = b_1(x - vt).
\tag{9.5}
$$

An event that occurs at $x = 0$, $t = 0$ necessarily occurs at $t' = 0$. Then the t' equation becomes

$$0 = c_2 y + c_3 z$$

for constants c_2 and c_3 that may depend on v.

This must be identically satisfied for all y and z. Hence, $c_2 = 0 = c_3$, yielding

$$t' = c_1 x + c_4 t \qquad (9.6)$$

for constants c_1 and c_4.

In fact, we can obtain the remaining spatial and temporal transformation equations directly from (9.5) without the assistance of (9.6).

Now, suppose the inertial frames S and S' reverse their direction such that they are both travelling in the $-x$ and $-x'$ directions, where S is now travelling at velocity $-v$ relative to S'. Performing the same analysis as above, we would obtain, instead of (9.5),

$$x = b_1 (x' + vt'), \qquad (9.7)$$

where b_1 would depend on $-v$. Remember that in obtaining (9.7), we have merely reversed the inertial frames. Thus the symmetry in going from (9.5) to (9.7) must be *even*,[6] which implies that $b_1(v) = b_1(-v) \equiv \gamma(v)$.[7] So in terms of γ, equations (9.5) and (9.7) become

$$x' = \gamma(v)(x - vt) \qquad (9.8)$$

and

$$x = \gamma(v)(x' + vt'). \qquad (9.9)$$

From Postulate 2 the speed of light c must be the same in all inertial frames. Thus the trajectory of a beam of light in S and S' is given, respectively, by

$$x = ct$$

and

$$x' = ct'.$$

Substituting these trajectories into (9.8) and (9.9) yields

$$ct' = \gamma(v)t(c - v)$$

and

$$ct = \gamma(v)t'(c + v).$$

---

[6]A function $f(x)$ is said to be *even* if $f(x) = f(-x)$. Correspondingly, the function $y(x)$ is said to be *odd* if $g(x) = -g(-x)$.

[7]γ is traditionally used to represent this coefficient; but it will also be used to represent a photon although no confusion should arise.

On performing some simple algebra, we find that

$$\gamma(v) = \sqrt{\frac{1}{1 - \beta^2}}, \qquad (9.10)$$

where

$$\beta \equiv \frac{v}{c}.$$

To obtain the temporal transformation, we simply substitute (9.8) into (9.9) to eliminate x', yielding

$$t' = \gamma(v)\left(t - \frac{\beta}{c}x\right). \qquad (9.11)$$

Similarly, on eliminating x in (9.8) and (9.9),

$$t = \gamma(v)\left(t + \frac{\beta}{c}x'\right). \qquad (9.12)$$

On combining (9.3), (9.4), (9.8) and (9.12), the full set of *Lorentz transformations* are

$$\begin{aligned}
x' &= \gamma(x - \beta ct) \\
y' &= y \\
z' &= z \\
ct' &= \gamma(ct - \beta x).
\end{aligned} \qquad (9.13)$$

Notice that the temporal transformation is given for ct' instead of just t'. This is so that each coordinate transformation has the same dimension — that of distance. Also, we have dropped the v dependency on γ. Unless there is a possibility of ambiguity — for example, in a case where two distinct particle velocities are discussed — the v dependency will be omitted.

The *inverse Lorentz transformations* are given by

$$\begin{aligned}
x &= \gamma(x' + \beta ct') \\
y &= y' \\
z &= z' \\
ct &= \gamma(\beta x' + ct').
\end{aligned} \qquad (9.14)$$

Notice, for the Lorentz transformations (9.13), when $\beta \to 0$, that is $v \ll c$, we have $\gamma \to 1$ and the Galilean transformations (1.4) are recovered. Also, (9.14) can be obtained from (9.13) by allowing $\beta \to -\beta$ and swapping primed and unprimed indices. This is allowed because of the principle of relativity.

Strictly speaking, the transformations (9.13) are more accurately described as *Lorentz boosts* for the same reasons given in Section 1.1.4 for the Galilean transformations. However, we will continue to use 'Lorentz transformations' for the transformations (9.13).

9.4 Minkowski diagrams (space-time diagrams)

An extremely useful aid for discussing a wide variety of special relativistic scenarios is the *Minkowski diagram*, also referred to as a *space-time diagram*. This type of diagram was originally introduced into special relativity in 1907 by the German mathematician Hermann Minkowski (1864-1909). He was also responsible for introducing the four-dimensional interpretation of space and time. The space-time of special relativity is called *Minkowski space*.

We will consider two-dimensional Minkowski diagrams. These diagrams have two spatial dimensions suppressed, namely the y- and z-dimensions. The temporal axis will be constructed so as to have the dimension of distance; that is, t is multiplied by c, the speed of light, then the vertical axis is the ct-axis. The other axis corresponds to displacements in the x-direction. Note that these axes are orthogonal to each other in $\mathbb{E}^2$.

The coordinates (ct, x) representing a point in an inertial frame S must be expressed in coordinates (ct', x') in an inertial frame S'. Somehow both frames, and therefore both coordinate systems, need to be represented in the same diagram if we are to use it to represent relative motions.

In Figure 9.3 the Minkowski diagram depicts two sets of axes: the ct- and x-axes in S and the ct'- and x'-axes in S'. Notice that the axes in S' are not parallel with those in S, although the origins of S and S' coincide, and if this occurs at $t = 0 = t'$ the relation between coordinates in S and S' is given by the Lorentz transformations. We will soon see that this configuration of the axes means that S' is travelling at a uniform speed v, say, relative to S. The point P in the diagram represents what is called an *event* in space-time. The coordinates of P in S are (ct_P, x_P), and in S' they are (ct'_P, x'_P).

Now, the ct'-axis is characterised by $x' = 0$. Bearing in mind that the y- and z-dimensions are being suppressed, we have, using the inverse Lorentz transformations (9.14)

$$x = \gamma \beta ct'$$

and

$$ct = \gamma ct'.$$

On eliminating t'

$$ct = \frac{1}{\beta}x. \tag{9.15}$$

This is the equation of the line that represents the ct'-axis. Notice that $\beta < 1$, which corresponds to velocities less than c, so the angle between the ct-axis and the ct'-axis cannot exceed $45°$.

Similarly, the x'-axis is characterised by $ct' = 0$. Again, with the aid of (9.14), we arrive at

$$ct = \beta x.$$

This is the equation of the line that represents the x'-axis. As $\beta < 1$, the

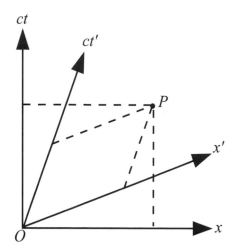

FIGURE 9.3
Minkowski diagram depicting an event P in S and S'

angle between the x-axis and the x'-axis cannot exceed 45°. Moreover, the angle $\widehat{ct0ct'}$ and $\widehat{x0x'}$ must be the same for a particular value of β and, hence, for a particular value of v, the velocity of S' relative to S.

It can be readily seen that as the axes in S' are brought together, i.e. the amount of tilting of the axes with respect to those in S, the greater the velocity of S' relative to S. The limiting case is when the ct'-axis and the x'-axis are parallel. This can never be achieved for massive particles (particles with mass), but will always be the case for photons and other massless particles.

9.4.1   Calibration hyperbolae and the s^2 invariant

As we delve farther into the depths of special relativity it will become ever more apparent that different observers occupying different inertial frames may have widely varying views with respect to a particular event in space-time. Space and time can no longer be considered to be absolute quantities, as was the case in the Newtonian scheme. Indeed, there are very few constants (or more accurately *invariants*) in special relativity, but those that exist are treated with at best a modicum of reverence.

Consider an event in S with coordinates (ct, x). An inertial observer in S' will record the event as occurring at (ct', x') in S'. What would the result be if we took the difference between the squared quantities $(ct')^2$ and $(x')^2$? With the aid of the Lorentz transformations (9.13), we will have

$$(ct')^2 - (x')^2 = \gamma^2(ct - \beta x)^2 - \gamma^2(x - \beta ct)^2$$
$$= \frac{(ct)^2 + (\beta x)^2 - x^2 - (\beta ct)^2}{1 - \beta^2}$$
$$= (ct)^2 - x^2$$

or

$$(ct')^2 - (x')^2 = (ct)^2 - x^2 = s^2, \tag{9.16}$$

where s^2 is an invariant for all inertial frames. If one defines the ct- and x-axes as orthogonal — as is usually the case — the second equality in (9.16) would define the hyperbolae

$$(ct)^2 - x^2 = \pm 1 \tag{9.17}$$

depicted in Figure 9.4. The hyperbolae (9.17) are called *calibration hyperbolae* because irrespective of which inertial frame is considered the intersection of

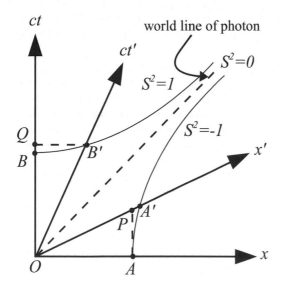

FIGURE 9.4
Calibration hyperbolae

$(ct)^2 - x^2 = -1$, say, with the x-axis of any inertial frame will define unit length in that frame. In particular, from Figure 9.4, $OA = 1 = OA'$. *Beware*, if you literally measured the distances OA and OA', you would of course find that $OA' > OA$. The Minkowski diagram in Figure 9.4 is merely a representation of the lengths in the inertial frames S and S'. Consequently, one must show a little caution when interpreting these diagrams. They are useful as an illustrative aid, but one must invariably rely upon analytical techniques to avoid possible contradictions.

The calibration hyperbola $s^2 = 1$ represents unit 'temporal' distance in the inertial frames S and S'; that is, $OB = 1 = OB'$.

The straight line $s^2 = 0$ is asymptotic to both hyperbolae $s^2 = 1$ and $s^2 = -1$. It represents the *world line* of a photon, i.e. the path that a photon will take in space-time after being emitted at the origin.

There are two important points that need to be emphasised with regard to Figure 9.4:

1. in S, let OA represent a rod of length l at $t = 0$. In S', the rod, travelling at velocity $-v$, is represented by the length OP at $t' = 0$. However, OP is less than OA' and as the calibration hyperbola intersects the x'-axis at $x' = 1$, the rod is less than 1 in S'. The implication is that moving rods are shorter according to an observer fixed in S;

2. in S', let the ct'-axis represent the world-line of a clock fixed at O in S', but moving with velocity v relative to S. At B', $ct' = 1$. However, the calibration hyperbola intersects the ct-axis at B, meaning that the point $ct = Q$ is greater than 1. The implication is that moving clocks tick slower according to an observer fixed in S. We will discuss the strange consequences of these two points in the next section.

Notice that (9.16) is independent of frame velocity. If S' were to be moving at a much faster speed $w > v$, say, the ct'-axis and x'-axis would be squeezed closer to the world line of the photon ($s^2 = 0$ in Figure 9.4). All this means is that A' and B' will be shifted farther along their respective calibration hyperbolae. Then an observer in S' will measure 1 metre and 1 second to be smaller relative to the observer in S.

9.5   Relativistic kinematics

From our discussion so far, and with regard to the Lorentz transformations in particular, it should be evident that our interpretation of events in special relativity runs contrary not only to the Newtonian interpretation, but also to our common sense.

9.5.1   Relativity of simultaneity

Events occurring at equal times t in S may not correspond to events occurring at equal times t' in S'. This is a consequence of the time transformation equation in the Lorentz transformations and is called *relativity of simultaneity*.

Recall from Newtonian dynamics that time is considered as absolute, so that two events P_1 and P_2 occur simultaneously if the times, $t_1 = t_2$, are the same. What does simultaneity mean in special relativity?

Consider the Minkowski diagram in Figure 9.5. Three events are depicted by points P_1, P_2 and P_3 with space-time coordinates (ct_1, x_1), (ct_2, x_2) and

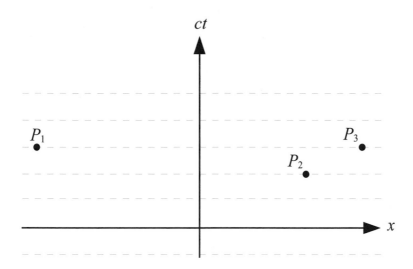

FIGURE 9.5

Lines of simultaneity in S

(ct_3, x_3), respectively, in the inertial frame S. An observer in S calls all events simultaneous for $t = $ constant. Thus each dashed line parallel to the x-axis is a *line of simultaneity*.

The events at P_1 and P_3 occur simultaneously because they are both on the same line of simultaneity, so $t_1 = t_3$. However, the event at P_2 lies on a different line of simultaneity, and does not occur simultaneously with events at P_1 and P_2.

This entire interpretation changes for an observer in S' travelling at a velocity v relative to S. This is depicted by the Minkowski diagram in Figure 9.6. Here, the same three events P_1, P_2 and P_3 have space-time coordinates (ct'_1, x'_1), (ct'_2, x'_2) and (ct'_3, x'_3), respectively. Now, P_2 and P_3 occur simultaneously because they are both on the same line of simultaneity, so $t'_2 = t'_3$, but P_1 does not occur on this line and the event is not simultaneous with the events at P_2 and P_3. Notice that the lines of simultaneity are, of course, parallel to the x'-axis.

Let us employ the Lorentz transformations to analyse the situation above. For simultaneous events P_1 and P_3 in S, we have $t_1 = t_3 = T$, a constant.

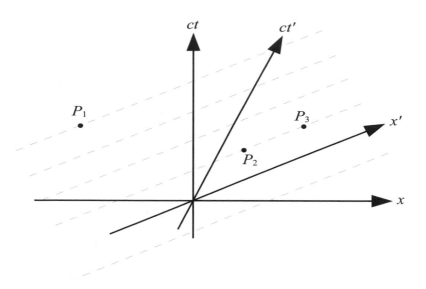

FIGURE 9.6
Lines of simultaneity in S'

Then from (9.13),
$$ct'_1 = \gamma(cT - \beta x_1)$$
and
$$ct'_3 = \gamma(cT - \beta x_3).$$

Subtracting,
$$c(t'_1 - t'_3) = \gamma\beta(x_3 - x_1).$$

Therefore, unless $x_1 = x_3$, the events in S' are not simultaneous. Furthermore, if we employ the inverse transformations then lines $t' = T' = $ constant are given by the equation
$$ct = \beta x + \frac{cT'}{\gamma},$$

which correspond to the lines of simultaneity in S' depicted in Figure 9.6.

The material point is that different observers will not, in general, agree about the time at which an event occurs. Einstein devised a *gedankenexperiment*[8] to illustrate the ideas above. The following example is an adaptation of this.

Example 9.1 A train travelling at speed v in the x-direction carries an observer positioned at O' (see Figure 9.7) in the frame S'. The train passes an observer at O in the stationary frame S. At time

---

[8]A gedankenexperiment comes form the German and means 'thought experiment'. Other notable examples are Schrödinger's cat and Maxwell's demon.

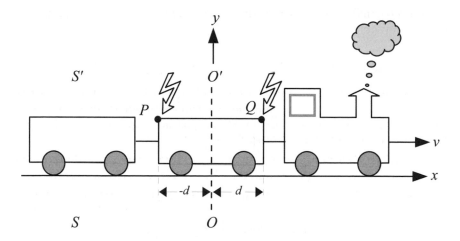

FIGURE 9.7
Einstein's gedankenexperiment to illustrate simultaneity

$t = 0$, in S, when O and O' are aligned along the y-axis, two bolts of lightning strike the carriages at P and Q, where $OP = -d$ and $OQ = d$. According to the observer's clock at O, both bolts strike the carriages simultaneously. How does the observer at O' interpret the events?

Solution At $t = 0$, in S, the observer at O sees these events at positions $x_P = -d$ and $x_Q = d$. However, in the S' frame the observer at O' records the positions $x'_P = -\gamma d$ and $x'_Q = \gamma d$. The times of these events, in S', are

$$t'_P = \frac{\gamma \beta d}{c}, \quad t'_Q = -\frac{\gamma \beta d}{c}.$$

This means that the observer at O' will see a bolt of lightning strike the carriage at P before the strike at Q — not simultaneous events. According to the observer's clock at O', the difference in time between strikes is

$$t'_P - t'_Q = \frac{2\gamma \beta d}{c}.$$

In practice, this difference will be exceedingly small: of the order 10^{-13} seconds for a conventional high speed train. $\qquad \square$

9.5.2 Time dilation

Consider a clock that is stationary in its rest frame S'. The frame S' is travelling at velocity v relative to an observer stationary in S. As measured in

S, the coordinates of two successive ticks of the clock (these represent two separate events) are

$$(ct_1, x_1), \quad (ct_2, x_2).$$

As measured in S', the coordinates of two successive ticks of the clock are

$$(ct'_1, x'_1), \quad (ct'_2, x'_2).$$

Because the clock is stationary in S', no spatial displacement takes place; therefore we can put $x'_1 = x'_2 \equiv x'_0$:

$$(ct'_1, x'_0), \quad (ct'_2, x'_0).$$

Let the interval between ticks in S and S' be given, respectively, by

$$\Delta t' = t'_2 - t'_1$$

and

$$\Delta t = t_2 - t_1.$$

Employing the temporal inverse Lorentz transformation (9.14), we have

$$ct_1 = \gamma(\beta x'_0 + ct'_1)$$

and

$$ct_2 = \gamma(\beta x'_0 + ct'_2).$$

On subtracting,

$$c(t_2 - t_1) = \gamma c(t'_2 - t'_1)$$

or

$$\Delta t = \gamma \Delta t'. \tag{9.18}$$

Hence, $\Delta t > \Delta t'$, unless $\gamma = 1$. The interval between ticks of the moving clock Δt is *greater* than the interval between ticks of the stationary clock $\Delta t'$ (remember that the clock is stationary in S'). The inference is that moving clocks tick more slowly than stationary clocks. The phenomenon is known as *time dilation*.

9.5.3 Experimental evidence of time dilation

There have, over the years, been a number of directly verifiable accounts of time dilation. In this section, we will merely mention the ideas of two of these experiments.

The first verifiable account was given by Rossi and Hall in 1940.[9] They were able to detect the presence of a particular unstable elementary particle, created in the upper layers of the atmosphere as a result of cosmic ray bombardment. This was witnessed at two separate altitudes: at Echo Lake in the Colorado

---

[9]See Rossi, B. and Hall, D.B. *Physical Review* **59**, 233 (1941).

Rocky Mountains (3230 m above sea level) and at Denver, Colorado (1616 m above sea level).

The particle in question is called a *muon* (mu-meson). They have a *half-life* (the time taken for half of a large number or particles to decay into other particles) of around 2×10^6 s. However, muons have been detected at sea level, which would take around 7×10^{-6} s to reach. As the speed of the muons is around 99 percent the speed of light, a quick calculation will show that in the muon's frame, its journey will only have taken

$$\frac{7 \times 10^{-6}}{\gamma} \approx 7 \times 10^{-7} \text{ s}.$$

Rossi and Hall calculated the number of muons that survived the journey from the upper atmosphere to the given locations above, at 3230 m and 1616 m, and concluded as we have done that the survival of these muons relies upon the fact that their clocks (in the moving frame) tick slower than the detectors' clocks (in the stationary frame). The number of muons detected at these levels were found to be consistent with time dilation effects.

A rather novel (although perhaps slightly unorthodox) approach to verifying time dilation was given by the Hafele–Keating experiment[10] in 1971, which utilised public transport! Two commercial airliners were flown around the world, one flying east and the other flying west. Stowed aboard each aircraft were two highly accurate cesium clocks. After the aircraft had completed their journeys their clocks were compared to another cesium clock located at the U.S. Naval Observatory in Washington, D.C. All clocks were synchronised prior to the commencement of the flights. It was found that the clocks travelling east gained (-59 ± 10) ns, while the clocks travelling west gained (273 ± 7) ns. This compares favourably with the predicted results: (-40 ± 23) ns flying east and (275 ± 21) ns flying west. The predicted results combined both special and general relativistic effects.[11] It should be pointed out that the reason for the large time difference between the east-west journeys is partly due to the rotation of the Earth and also the frames concerned are not quite inertial.

9.5.4 The twin paradox

Possibly the most contentious of all the gedankenexperiments of special relativity is the *twin paradox* (or *clock paradox*). From a commonsense perspective it will again seem absurd, although it is, of course, actually a well-founded statement of the theory. In this sense it will always remain a paradox. However, the actual paradox refers to a perceived contradiction of the basic tenets of the theory. We will presently see that no contradictions arise and, therefore, there is no paradox to resolve.

---

[10]Hafele, J.C. and Keating, R.E. *Science* **177**, 166 (1972).

[11]In general relativity (Einstein's theory of gravity) clocks tick more slowly as the gravitational field increases. Thus clocks at high altitude tick faster than those on the ground.

Let us begin with a story. Once upon a time there lived two twins: Arnie and Barnie. One day, Barnie decided to pop over and visit an old friend who lived on a very distant planet. Barnie set off in his spaceship at speed v to see his pal. Now at time T, in Arnie's frame, Arnie saw Barnie arrive on the distant planet. Satisfied that Barnie had arrived safely, Arnie put the kettle on for a cup of tea. Excited at the prospect of seeing his friend, Barnie slowed his spaceship down to land — but unfortunately, he had forgotten to bring the address with him and decided to turn back and go home, again travelling at speed v. When Barnie arrived back home, to his surprise Arnie had aged $T_A = 2T$, while Barnie had aged $T_B = 2T/\gamma$: Arnie was now older than Barnie! They both decided that the best course of action was to discuss this 'paradox' over a cup of tea.

On closer inspection it becomes obvious that all we have done is reiterate the time dilation gedankenexperiment, albeit in a rather fascinating tale. So where lies the paradox in this tale?

From Arnie's perspective there is no paradox. He knows that moving clocks tick slower and so is quite unsurprised to see his 'younger' twin brother bounding spritely in to greet him on his return. Barnie is surprised to find Arnie is not younger than himself. From Barnie's perspective it is Arnie who travelled at speed v in the opposite direction relative to his stationary spaceship. This is the paradox. According to the principle of relativity (Postulate 1): all inertial frames are equivalent with respect to the laws of physics. And as we all know, moving clocks tick slower. So why has nature decreed that Barnie should be the one to age less?

The simple answer to this question is that symmetry is broken between Arnie's journey and Barnie's journey. So why is this situation not symmetric? Well, do not forget that Barnie was slowing his spaceship (decelerating) so as to land on the planet, then turned his spaceship around and accelerated back up to speed v for the return journey, whereas in Arnie's frame no such deceleration or acceleration took place. Hence, Arnie's journey and Barnie's journey are not symmetric: Barnie's frame is no longer inertial while decelerating and accelerating. This scenario can be more easily visualised with the help of Minkowski diagrams.

First, consider Barnie's outward journey. Arnie's stay-at-home frame is S in Figure 9.8 while Barnie's inertial moving frame is S'. The position of the planet is H.

At the origin of S, Arnie has coordinates $(ct, x) = (0, 0)$. At the origin of S', Barnie has coordinates $(ct', x') = (0, 0)$. Before Barnie departs, the brothers synchronise their clocks. At H in the frame S

$$(ct, x) = (cT, vT).$$

At H in the frame S', we can determine the time at which Barnie arrives at

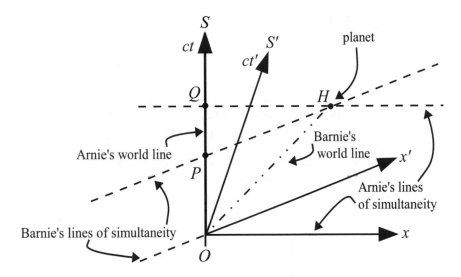

FIGURE 9.8

Barnie's outward journey

H (according to him). Thus, from (9.13)

$$ct' = \gamma(ct - \beta x)$$

$$t' = \gamma T(1 - \beta^2) = \frac{T}{\gamma}.$$

So if Arnie has aged $T_A = T$ years when Barnie reaches H, then Barnie has aged $T_B = T/\gamma$ years. Hence,

$$T_B = \frac{T_A}{\gamma} \tag{9.19}$$

implying that Barnie has aged by an amount $1/\gamma$ less than Arnie.

In Barnie's frame the point P lies on his line of simultaneity: this is the point where Barnie thinks Arnie is when Barnie reaches the planet at H. We can, therefore, use the temporal equation in (9.13) to determine by how much Arnie has aged compared with Barnie. (Note that $x = 0$ at P.) Thus

$$ct' = \gamma(ct - 0)$$

or

$$t = \frac{t'}{\gamma}.$$

So if Barnie has aged $T_B = T/\gamma$ years then Arnie has aged $T_A = (T/\gamma)/\gamma$ years. Hence

$$T_A = \frac{T_B}{\gamma} \tag{9.20}$$

implying that Arnie has aged by an amount $1/\gamma$ less than Barnie.

The times (9.19) and (9.20) are consistent with what one would expect from the effects of time dilation. Symmetry is preserved because both Arnie and Barnie have aged with respect to each other by the same factor.

Symmetry is broken, however, when Barnie decides to return home — the frame in which he decelerates and accelerates again is no longer inertial. Nevertheless, we can conceive of a situation where just before Barnie commences deceleration, a friend of Arnie's (Marnie) passes Barnie at speed v and travels in the opposite direction in her spaceship to visit Arnie. As they pass each other, both Barnie and Marnie synchronise their clocks. As Marnie's frame is inertial, symmetry is again preserved.

Let us now combine all the events together in a single Minkowski diagram (Figure 9.9).

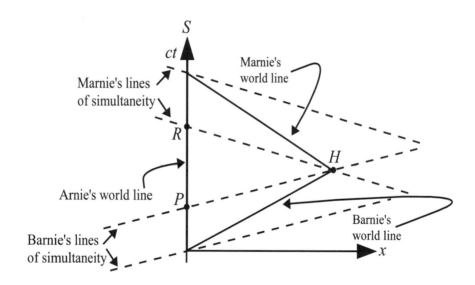

FIGURE 9.9
Marnie visits Arnie

Now from this diagram, we can see that as Barnie arrives at H he believes that Arnie is at P (both points lying along Barnie's line of simultaneity). However, as Marnie passes Barnie at point H, she believes that Arnie is at point R (both points lying along Marnie's line of simultaneity). Clearly, Barnie and Marnie are in complete disagreement as to Arnie's age at point H where they synchronise clocks.

The discrepancy arises because Barnie and Marnie are travelling in opposite directions. Indeed, had Barnie reversed direction as he reached H then in the time taken for him to complete the manoeuvre, Arnie would have aged sufficiently for Barnie to determine Arnie's point at R.

9.5.5 The Lorentz–Fitzgerald contraction (length contraction)

Figure 9.10 depicts a Minkowski diagram for a rod in two frames: $S(ct, x)$ and $S'(ct', x')$. As usual, S' is moving at velocity v relative to S. The rod, station-

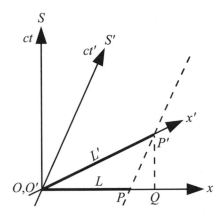

FIGURE 9.10
Length contraction of a moving rod

ary in S', has length L' and lies along the x'-axis between two simultaneous end points O' and P'. Thus, the coordinates of the end points in S' are

$$O'(ct', x') = (0, 0)$$
$$P'(ct', x') = (0, L').$$

(Note that it is vitally important that the coordinates are specified for the same times; otherwise length cannot be defined in this manner.)

According to an observer in S the rod is moving at velocity v. The coordinates of the end points of the moving rod in S can be determined with the aid of the inverse Lorentz transformations (9.14). Thus, for end point O'

$$x = 0$$

and

$$ct = 0;$$

for end point P'

$$x = \gamma(x' + \beta ct') = \gamma L'$$

and

$$ct = \gamma(\beta x' + ct') = \beta \gamma L'.$$

So, the coordinates of the end points in S are

$$O' = (ct, x) = (0, 0)$$

and

$$P'(ct, x) = (\beta\gamma L', \beta L').$$

However, these coordinates will not allow us to determine the *length* of the rod in S. The reason for this is because O' and P' are not the coordinates of simultaneous events in S. What one must do is project the end point at P' along its world line to a new end point at P in S. Now O and P are simultaneous events in S.

The coordinates of O are the same as that of O':

$$O(ct, x) = (0, 0).$$

The coordinates of P in S are

$$\begin{aligned} P(ct, x) &= (O, \gamma L' - PQ) \\ &= (O, \gamma L' - vt) \\ &= (O, \gamma L' - \beta^2 \gamma L') \\ &= (O, L'/\gamma). \end{aligned}$$

But the coordinates of P in S are also

$$P(ct, x) = (O, L).$$

Thence, the length of the rod in S is

$$L = \frac{L'}{\gamma}. \tag{9.21}$$

This means that the moving rod as seen by an observer in S is *contracted* by a factor of $1/\gamma$ along its direction of motion; we call this phenomenon *Lorentz–Fitzgerald contraction* (compare with equation (9.1)). The length of the rod L' fixed in its moving frame S' is sometimes called the *proper length*. As was mentioned in Section 9.1.1 some caution should be taken when interpreting the Minkowski diagrams. Notice in Figure 9.10 the length of the rod in the moving frame has been drawn *longer* than in the stationary frame. Again, we reiterate: Minkowski diagrams are useful visualisations of events, but they do not always depict an accurate geometrical representation of these events.

Example 9.2 The pole paradox Clark could be the greatest sprinter in history. He regularly runs at a speed $(\sqrt{3}/2)c$ ($\gamma = 2$) but he refuses to do so unless he is allowed to carry his favourite 20-m pole! Clark's mother, Mrs. Kent, enjoys watching him sprint through their 10-m barn while she is milking the cow. But one particular day, she notices that the back door of the barn is closed with only a small aperture wide enough for the pole to protrude through. Before she is able to warn Clark, he comes thundering in, clutching the horizontal 20-m pole. However, to Mrs. Kent's delight, the pole requires no aperture to pass through as she observes it completely fitting into the barn. However, Clark observes that only 5 m fits within the barn and the remaining 15 m protrudes out of the back of the barn through the aperture. How can this 'paradox' be resolved?

Solution The Minkowski diagram Figure 9.11 depicts this scenario from both Clark's and Mrs. Kent's perspectives! Seen from Mrs. Kent's perspective (the S frame), the 20-m pole 'fits' inside the barn

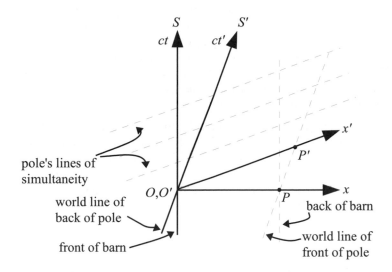

FIGURE 9.11
The pole and the barn seen from two different perspectives

easily. Here, by 'fit', we mean that the events at O (back of pole is within barn) and P (front of pole is within barn) are simultaneous in S. These two events are simultaneous in S' — Clark's perspective. Instead, the events at O' (back of pole is within barn) and P' (front of pole protrudes from back of barn) are simultaneous in S' — the pole does not 'fit' within the barn.

Indeed, we need only appeal to (9.21) to verify the length contraction process from the point of view of each observer. Thus, in Mrs. Kent's frame

$$10 = \frac{20}{2}$$

and in Clark's frame

$$5 = \frac{10}{2}.$$

What would have transpired had there been no aperture for the pole to slide through? From Mrs. Kent's perspective all is well until the pole hits the back of the barn, but that will not occur until the pole is completely within the barn. On the other hand, the pole will impact on the back door of the barn before Clark has even entered (assuming he is holding the pole somewhere equidistant from either end). But, of course, he will not be aware of this immediately, for the information relating to this event will take a finite time to reach Clark. Similarly, the information, in the form of a shockwave propagating along the pole, will take a finite time to reach the other end of the pole. This means that the back end of the pole will remain stationary while the barn is engulfing it. Indeed, had the barn been travelling at $(3/4)c$, the information that the front of the pole had suffered an impact would reach the back end of the pole just as the barn swallowed it up. But, from Clark's perspective, the barn is travelling at $(\sqrt{3}/2)c$, so the end of the pole is within the barn before it has received the information that the front end has impacted on the back of the barn door.

Thus each observer in their inertial frame will necessarily view events differently — each observer has their own story to tell from their perspective. Moreover, each story is a valid representation of events, no matter how different each one is, provided the frames are inertial and the physics that characterises these events remains invariant under Lorentz transformations.   □

9.5.6   Transformation of velocity

So far, we have considered only inertial frames in which objects are stationary within frames, but the frames themselves are moving relative to each other. We must now consider the situation where an object is *moving* within a frame and that frame is moving relative to another frame. For example, a frame S', moving at velocity v relative to the frame S, has an object moving at velocity u. What is the velocity u of the object in S? To answer this question, we need a formula that will allow us to transform between u' and u. In the Galileo–Newton formulation it is merely (1.2), i.e. $u' = u - v$. However, this is clearly not so in special relativity; for if $u' = c'$ (the speed of light in S') and $u = c$ then $c' \neq c - v$.

Consider a particle with position vector $\mathbf{r}(t) = (x(t), y(t), z(t))$ in S and $\mathbf{r}'(t') = (x'(t'), y'(t'), z'(t'))$ in S'. The particle's velocity $\mathbf{u} = (u_x, u_y, u_z)$ in S is

$$(u_x, u_y, u_z) = \left(\frac{dx}{dt}, \frac{dy}{dt}, \frac{dz}{dt} \right), \tag{9.22}$$

and the velocity $\mathbf{u'} = (u'_x, u'_y, u'_z)$ in S' is

$$(u'_x, u'_y, u'_z) = \left(\frac{dx'}{dt'}, \frac{dy'}{dt'}, \frac{dz'}{dt'}\right). \tag{9.23}$$

To transform between the velocity components of $\mathbf{u'}$ and $\mathbf{u}$ requires *taking differentials*[12] of the Lorentz transformations (9.13):

$$dx' = \gamma(dx - \beta c dt) \tag{9.24}$$
$$dy' = dy \tag{9.25}$$
$$dz' = dz \tag{9.26}$$
$$cdt' = \gamma(cdt - \beta dx). \tag{9.27}$$

So, the velocity transformation for the first component of (9.23) is

$$
\begin{aligned}
u'_x = \frac{dx'}{dt'} &= \frac{c\gamma(dx - \beta c dt)}{\gamma(cdt - \beta dx)} \quad \text{from (9.24) and (9.27)} \\
&= \frac{c(dx/dt - \beta c)}{c - \beta dx/dt} \quad \text{on factoring } dt \\
&= \frac{u_x - \beta c}{1 - (\beta/c)u_x} \quad \text{from (9.23).}
\end{aligned}
$$

On following similar steps to obtain u'_y and u'_z, the full set of *velocity transformation formulae* are

$$u'_x = \frac{u_x - \beta c}{1 - (\beta/c)u_x}, \quad u'_y = \frac{u_y}{\gamma(1 - (\beta/c)u_x)}, \quad u'_z = \frac{u_z}{\gamma(1 - (\beta/c)u_x)}. \tag{9.28}$$

The inverse velocity transformation formulae can be obtained in a straightforward fashion by letting $\beta \to -\beta$ ($v \to -v$) and $\mathbf{u} \leftrightarrow \mathbf{u'}$. Thence,

$$u_x = \frac{u'_x + \beta c}{1 + (\beta/c)u'_x}, \quad u_y = \frac{u'_y}{\gamma(1 + (\beta/c)u'_x)}, \quad u_z = \frac{u'_z}{\gamma(1 + (\beta/c)u'_x)}. \tag{9.29}$$

For velocities much less than c, that is, for $\beta \to 0$ and $\gamma \to 1$, we recover the Galilean velocity transformation formulae (see (1.6))

$$u'_x = u_x - v, \quad u'_y = u_y, \quad u'_z = u_z. \tag{9.30}$$

Assuming that the frame and particle velocities are co-linear in the x-direction then $\mathbf{u} = (u_x, 0, 0)$ and, of course, $\mathbf{u'} = (u'_x, 0, 0)$. So the velocity transformation formulae (9.28) and (9.29) reduce, respectively, to

$$u' = \frac{u - \beta c}{1 - (\beta/c)u} \tag{9.31}$$

---

[12]If $y = f(x)$ where df/dx is the derivative of $f(x)$, then on taking differentials: $dy = \frac{df}{dx}dx$.

and

$$u = \frac{u' + \beta c}{1 + (\beta/c)u'}. \tag{9.32}$$

Notice that for a *light-like particle*(particle travelling at the speed of light) moving in S, that is, $u = c$, the velocity of this particle in S' is $u' = c$, which is what would be expected. It can also be shown that u' will always be less than c regardless of how close u and v are to c, that is, for $|u| < c$ and $|v| < c$. Thus

$$c - u' = \frac{c - \beta u - u + \beta c}{1 - (\beta/c)u} = \frac{c(c - u)(1 - \beta)}{c - \beta u} > 0,$$

and as $\beta < 1$, the last inequality holds implying that $u' < c$. One can show similarly that for $|v| < c$, $u' > -c$ (see Exercise 9.10).

> **Example 9.3** The starship Nebuchadnezzar, fleeing from the space station Balthazar, immediately comes under attack. As the Nebuchadnezzar reaches 0.95 of light speed relative to the Balthazar, it receives a direct hit. The crew just manage to escape into their shuttlecraft Salmanazar before the Nebuchadnezzer is destroyed. If the Salmanazar leaves the Nebuchadnezzar with relative speed $0.75c$ in the same direction as the Nebuchadnezzar, what is its speed relative to the Balthazar? Compare this result with one obtained by using the Galilean velocity transformation formulae (9.30). (Assume all motion is co-linear.)
>
> **Solution** Assume that the Balthazar is our frame S and the Nebuchadnezzar is the S' frame travelling at $v = 0.95c$, then $u' = 0.75$ for the Salmanazar. Substituting these values into (9.32) gives
>
> $$u = \frac{(0.75 + 0.95)c}{1 + (0.95)(0.75)} = \frac{1.7c}{1.7125} \approx 0.99c,$$
>
> which is the speed of the Salmanazar relative to the Balthazar. Comparing this with the value given by (9.30):
>
> $$u = (0.75 + 0.95) = 1.7c$$
>
> shows how significant the difference can be between relativistic and classical Newtonian results. Moreover, when combining speeds relativistically, no matter how close to the speed of light they may be, the 'sum' will always be less than c. □

The velocity transformation formulae can be recast in a rather elegant form by introducing a new parameter called the *rapidity* ρ. To see this, consider the Lorentz transformations (9.13) and form the combinations $x' - ct'$ and $x' + ct'$:

$$x' - ct' = e^\rho(x - ct) \tag{9.33}$$

and

$$x' + ct' = e^{-\rho}(x + ct), \tag{9.34}$$

where e^ρ is defined as

$$e^\rho = \gamma(1 + \beta). \tag{9.35}$$

Note that the product of (9.33) and (9.34) produces the invariant (9.16).

It is now a simple task to obtain the following hyperbolic relations using (9.35):

$$\sinh\rho = \gamma, \quad \cosh\rho = \beta\gamma, \quad \tanh\rho = \beta. \tag{9.36}$$

Now, on dividing (9.31) by c, we have

$$\frac{u'}{c} = \left(\frac{u}{c} - \beta(v)\right)\bigg/\left(1 - \beta(v)\frac{u}{c}\right)$$

or

$$\beta(u') = \frac{\beta(u) - \beta(v)}{1 - \beta(u)\beta(v)}.$$

On incorporating (9.36), this becomes

$$\tanh\rho(u') = \frac{\tanh\rho(u) - \tanh\rho(v)}{1 - \tanh\rho(u) + \rho(v)}$$
$$= \tanh(\rho(u) + \rho(v)).$$

Hence, the velocity transformation formula can now be written as

$$\rho(u') = \rho(u) - \rho(v). \tag{9.37}$$

9.6 Exercises

1. An observer in an inertial frame witnesses a moving clock lose 1 second every 24 hours. Determine the speed of the clock relative to the observer.

2. Arnie wakes up one morning with a sudden desire to visit his pal Marnie who lives on a planet far, far away. On telling his twin brother, Barnie, to expect him back for tea, Arnie blasts off in his spaceship at $0.95c$ to visit Marnie. However, on arriving at Marnie's house he is disappointed to find that she has gone away on holiday. Arnie blasts off in his spaceship again at $0.95c$ to return home. On arriving back home, Barnie protests that not only is the tea cold, but he has been away for 80 years. How long does Arnie think he has been away for?

3. The half-life of a particular elementary particle called a *pion* is $t_h = 1.77 \times 10^{-8}$ s in its rest frame. Pions accelerated to a speed of $0.99c$ are fired at a detector 39 m away from the accelerator. If half of these pions are detected, what is the half-life of a pion in the observer's frame? What distance will the accelerator have covered, according to an observer in the pion's rest frame, during the pion's half-life t_h?

4. The velocity of the inertial frame S' relative to the inertial frame S is v. Two events are observed at a distance $x = l$ apart in S and a distance $x' = l'$ apart in S'. If the frames are in standard configuration and the events in S are observed simultaneously, show that

$$v = \frac{c}{l'^2}\sqrt{l'^2 - l^2}.$$

Show also that the time interval between the two events as measured in S' is $\sqrt{l'^2 - l^2}/c$.

5. Arnie realises that he is late for tea again and rushes in his spaceship to Barnie, who is waiting for him on Earth. However, in his haste, Arnie rushes past the Earth at speed $c/2$. Fortunately, Barnie is watching Arnie through his telescope and signals to Arnie with his laser light beam 1 second after Arnie passes Earth. How long will it take for Arnie to receive Barnie's signal according to Barnie, and according to Arnie?

6. Two observers Q and Q' are at rest at the spatial origins of inertial frames S and S', respectively, where S' is moving at speed v relative to S. At the point where the spatial origins S and S' coincide, Q and Q' have times $t = 0$ and $t' = 0$, respectively. Draw a Minkowski diagram to represent the following scenario: at time T'_2 in S', Q' receives a light signal sent by Q at time T_1 in S.

   Show that Q' receives the light signal while at coordinates $(\gamma c T'_2, \gamma v T'_2)$ in S. Hence, find a relation between the times T_1 and T'_2 in terms of β.

7. Show that the Lorentz transformations (omitting y' and z') can be written in hyperbolic form as

$$x' = x \cosh\rho - ct \sinh\rho \qquad (9.38)$$
$$ct' = ct \cosh\rho - x \sinh\rho. \qquad (9.39)$$

8. Consider two inertial frames S and S' in standard configuration. Let the instantaneous velocity of a particle in S be given by $\mathbf{u} = (u_x, u_y, u_z)$ and in S' be given by $\mathbf{u}' = (u'_x, u'_y, u'_z)$.

   If the acceleration of the particle in S is given by $\mathbf{a} = (a_x, a_y, a_z)$ and in S' is given by $\mathbf{a}' = (a'_x, a'_y, a'_z)$ where, by definition, $a_i = \frac{du_i}{dt}$

and $a'_i = \frac{du'_i}{dt'}$ for $i = x, y, z$, show using the velocity transformation formulae that the transformation of acceleration between S' and S is given by

$$a'_x = \frac{\gamma^{-3} a_x}{\alpha^3}$$

$$a'_y = \frac{\gamma^{-2}}{\alpha^2}\left(a_y + \frac{(\beta/c)u_y a_x}{\alpha}\right)$$

$$a'_z = \frac{\gamma^{-2}}{\alpha^2}\left(a_z + \frac{(\beta/c)u_z a_x}{\alpha}\right),$$

where

$$\alpha = 1 - (\beta/c)u_x.$$

9. By allowing $t \to it$ and $t' \to it'$ $(i = \sqrt{-1})$ in (9.38) and (9.39) show that they represent an anticlockwise rotation in the (ct, x)-plane through an angle ϕ, where ϕ is pure imaginary.

10. Consider two inertial frames S and S' with relative speed $|v| < c$. Show that the velocity u' of a particle in S' is greater than $-c$.

10

Space-time

CONTENTS

In this chapter, we will be specifically discussing a construct called *space-time*. More precisely, in our context the space-time under consideration is 4-dimensional Minkowski space-time, which is now more commonly referred to as *Minkowski space* $\mathbb{M}^4$.

The Einsteinian view of the universe embodied in special relativity as described in Chapter 9 loses none of its integrity and, indeed, is enhanced by its treatment within the framework of space-time. The power and elegance that are rightly attributed to the concept of space-time stem (as is usually the case in physics) from its simplicity. The fact that time can now be thought of as a mere fourth coordinate within the fabric of space-time will make it possible to develop a geometry analogous to Euclidean geometry $\mathbb{E}^3$. Thence, the familiar *Euclidean metric* of the Galileo–Newton formulation

$$dr^2 = dx^2 + dy^2 + dz^2 \qquad (10.1)$$

is replaced by the *Minkowski metric* in the Einstein formulation

$$ds^2 = c^2 dt^2 - dx^2 - dy^2 - dz^2, \qquad (10.2)$$

where ds^2 is the *infinitesimal invariant interval*.

A statement, oft quoted but pertinent, made by Minkowski in the light of his new discovery is: "...space by itself, and time by itself, are doomed to fade away into mere shadows, and only a kind of union of the two will preserve an independent reality." This is more true now than it has ever been.

10.1   The light cone

Consider a $(1+1)$-dimensional[1] space-time as defined by the *invariant interval* Δs^2:

$$\Delta s^2 = c^2 \Delta t^2 - \Delta x^2, \qquad (10.3)$$

where for two events $P_1(ct_1, x_1)$ and $P_2(ct_2, x_2)$ in an inertial frame S, the temporal and spatial separations of the events are given, respectively, by

$$\Delta t = t_2 - t_1, \quad \Delta x = x_2 - x_1. \qquad (10.4)$$

Notice the analogy between (10.3) and (9.16). The invariance of Δs^2 can be established in a similar manner. Thus, in S', on substituting the Lorentz transformations (9.14) into (10.3), we have

$$\begin{aligned}
\Delta s^2 &= \gamma^2[(\beta\Delta x' + c\Delta t')^2 - (\Delta x' + \beta c\Delta t')^2] \\
&= \gamma^2(1 - \beta^2)(c^2\Delta t'^2 - \Delta x'^2) \qquad (10.5) \\
&= c^2\Delta t'^2 - \Delta x'^2.
\end{aligned}$$

So, on combining (10.3) and (10.5) the analogous expression for (9.16) is

$$\Delta s^2 = c^2\Delta t'^2 - \Delta x'^2 = c^2\Delta t^2 - \Delta x^2. \qquad (10.6)$$

Now without loss of generality choose event P_1 at $(0,0)$ and event P_2 at (ct, x). There are three possible ranges for Δs^2: $\Delta s^2 > 0$, $\Delta s^2 = 0$ and $\Delta s^2 < 0$. Figure 10.1 depicts these three possibilities.

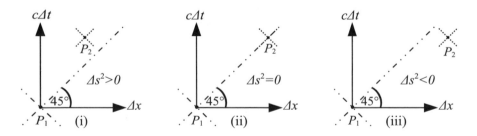

FIGURE 10.1
(i) time-like, (ii) light-like and (iii) space-like separated intervals

Although Figure 10.1 clearly depicts Minkowski diagrams, we will here consider them as representing *(1+1)-dimensional light cones* (dotted and dotted-dashed lines). That is, P_1 and P_2 are at the origins of their respective $(1+1)$-dimensional light cones. The dotted lines emanating from the origins of these

---

[1]$(1 + 1)$-dimensional corresponds to one time coordinate combined with one space coordinate.

light cones are the world lines of photons and must be at an angle of 45° with respect to the coordinate axes.

In (i) P_2 is within the future light cone of P_1. Conversely, P_1 is in the past light cone of P_2. So a signal sent by P_1 travelling at a speed less than c will be detected by P_2. The separation of these events is given for $\Delta s^2 > 0$ in (10.6) and we refer to them as *time-like separated*.

In (ii) P_2 lies along the edge of the light cone of P_1. Conversely, P_1 lies along the edge of the light cone of P_2. So only a signal sent by P_1 travelling at the speed of light will be detected by P_2. The separation of these events is given for $\Delta s^2 = 0$ and we refer to them as *light-like* (or *null*) *separated*.

In (iii) P_2 lies outside the light cone of P_1. Conversely, P_1 lies outside the light cone of P_2. So any signal sent by P_1 travelling at the speed of light or less will never be detected by P_2. The separation of these events is given for $\Delta s^2 < 0$ and we refer to them as *space-like separated*.[2]

In the case of a $(1 + 2)$-dimensional Minkowski space, the $(1 + 1)$-dimensional light cone is replaced by the conventional $(1 + 2)$-dimensional version as depicted in Figure 10.2.

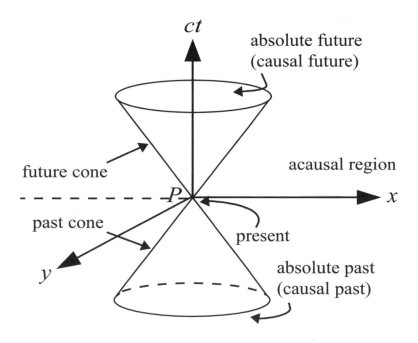

FIGURE 10.2
The (1+2)-dimensional light cone

---

[2] In 1967 it was postulated by Gerald Feinberg (1933-1992) that a particle known as a *tachyon*, possessing imaginary mass ($m^2 < 0$), could travel at speeds always greater than c. However, tachyons have been ruled out because of their unphysical nature in other areas of physics.

Consider yourself at event P. Any event that is within your *future cone*, or on its surface, can be influenced by you. In other words, any signal transmitted by you at speeds $v \leq c$ will be received by an event on or within your future cone. In addition, any observer positioned along the world line of the signal will agree as to the chronology of events: all other events in this region occur *after* you send the signal. Events in this region are *causally connected*, and form part of your absolute future.

Any event that is within your past cone, or on its surface, cannot be influenced by you. However, they will have influenced you. Events in this region constitute your *absolute past*, and again are causally connected.

There is a third region: the *acausal region* (or *elsewhere*). Events in this region neither affect nor are affected by you. For signals to reach events in this region, they would have to travel faster than the speed of light; that is, cross the light barrier, which is forbidden in special relativity. Events in this region are not causally connected to events within or on the light cone.

10.2   Proper time

Consider the world line of a particle given by the trajectory $x = x(t)$ in one spatial dimension. For the particle at rest at the origin ($x = 0$) in S, the invariant interval Δs^2 between two separate events on the particle's world line is, from (10.3),

$$\Delta s^2 = c^2 \Delta t^2.$$

Now, both Δs^2 and c^2 are invariants. The time interval between the events must also be invariant for all inertial frames. This time is known as the *proper time* τ, and we write it as

$$\Delta \tau = \frac{\Delta s}{c}.$$

Thus, in the rest frame

$$\Delta \tau = \Delta t.$$

If the points are infinitesimally separated and the particle moves at velocity $u = dx/dt$ along the trajectory, then

$$d\tau^2 = \frac{ds^2}{c^2} = dt^2 \left(1 - \frac{1}{c^2} \frac{dx^2}{dt^2} \right)$$

or

$$\frac{d\tau^2}{dt^2} = \left(1 - \frac{u^2}{c^2} \right)$$

implying that

$$\frac{dt}{d\tau} = \gamma(u), \tag{10.7}$$

where the positive root has been taken to ensure that $t, \tau > 0$.

10.3   The four-component vector formalism

The four-component vector formalism (or *4-vector formalism*) is in many respects a fundamental notion in special relativity. One may consider the 4-vector as the four-dimensional analogue of 3-vectors in dynamics, with one very important difference: time is now incorporated, so when combined with c it forms a fourth spatial component. We have already met this idea in two-dimensions and *via* (10.2).

The space in which 4-vectors reside is Minkowski space $\mathbb{M}^4$. The position 4-vector in $\mathbb{M}^4$ has components

$$X = (ct, x, y, z). \tag{10.8}$$

Notice that we defined the 4-vector by a non-bold uppercase letter X. This is so that we can distinguish between 4-vectors and bold lowercase 3-vectors. A conscious decision has been made not to represent 4-vectors in bold type. Thus no confusion should arise between 4-vectors and other non-bold quantities.

It is also sometimes convenient to write (10.8) as

$$X = (ct, \mathbf{x}), \tag{10.9}$$

where $\mathbf{x} = (x, y, z)$ is the position 3-vector. This will be utilised as appropriate.

The transformation of coordinates between inertial frames is achieved by employing the Lorentz transformations L. That is, for $X \in S$ and $X' \in S'$, we have

$$X' = LX \tag{10.10}$$

provided that the matrix L satisfies the equation

$$L^T \eta L = \eta, \tag{10.11}$$

where η is called the *Minkowski metric*:

$$\eta = \begin{pmatrix} 1 & 0 & 0 & 0 \\ 0 & -1 & 0 & 0 \\ 0 & 0 & -1 & 0 \\ 0 & 0 & 0 & -1 \end{pmatrix}. \tag{10.12}$$

Equation (10.11) is satisfied by L if it takes the form

$$L = \begin{pmatrix} 1 & 0 & 0 & 0 \\ 0 & R_{11} & R_{12} & R_{13} \\ 0 & R_{21} & R_{22} & R_{23} \\ 0 & R_{31} & R_{32} & R_{33} \end{pmatrix} \tag{10.13}$$

or

$$L = \begin{pmatrix} \gamma & -\beta\gamma & 0 & 0 \\ -\beta\gamma & \gamma & 0 & 0 \\ 0 & 0 & 1 & 0 \\ 0 & 0 & 0 & 1 \end{pmatrix}. \tag{10.14}$$

The matrix L in (10.13) can be seen to have no effect on the transformation of time between inertial frames. However, the 3×3 sub-matrix is a rotation matrix[3] R:

$$R = \begin{pmatrix} R_{11} & R_{12} & R_{13} \\ R_{21} & R_{22} & R_{23} \\ R_{31} & R_{32} & R_{33} \end{pmatrix} \qquad (10.15)$$

satisfying

$$R^T R = I,$$

where I is the 3×3 identity matrix (see line below (1.7)). Three separate rotations are incorporated in (10.15). Each can be parameterised by a different angle and so correspond to rotations about three spatial axes. However, rotations will not play a part in our discussion.

The matrix L in (10.14) corresponds to a boost along the x-axis with y and z suppressed. In other words, (10.14) is the matrix form of the Lorentz transformations (9.13):

$$\begin{pmatrix} ct' \\ x' \\ y' \\ z' \end{pmatrix} = \begin{pmatrix} \gamma & -\beta\gamma & 0 & 0 \\ -\beta\gamma & \gamma & 0 & 0 \\ 0 & 0 & 1 & 0 \\ 0 & 0 & 0 & 1 \end{pmatrix} \begin{pmatrix} ct \\ x \\ y \\ z \end{pmatrix}, \qquad (10.16)$$

which is just (10.10).

This is a particularly useful way of remembering the form of the Lorentz transformations. If we are not concerned with motion along the y- and z-axes, they can be reduced to the simple form

$$\begin{pmatrix} ct' \\ x' \end{pmatrix} = \begin{pmatrix} \gamma & -\beta\gamma \\ -\beta\gamma & \gamma \end{pmatrix} \begin{pmatrix} ct \\ x \end{pmatrix}. \qquad (10.17)$$

10.3.1   Lorentz invariance of the inner product

Consider two arbitrary 4-vectors U and V. The *inner product* of these two vectors is defined as:

$$(U)^{\mathrm{T}} \eta(V) \equiv U \cdot V,$$

where $U = (U^0, U^1, U^2, U^3)$ and $V = (V^0, V^1, V^2, V^3)$. Then

$$U \cdot V = U^0 V^0 - U^1 V^1 - U^2 V^2 - U^3 V^3 - U^4 V^4$$
$$= U^0 V^0 - \mathbf{u} \cdot \mathbf{v},$$

where $\mathbf{u} \cdot \mathbf{v}$ is the conventional inner product for 3-vectors.

---

[3]It could also represent a reflection for $\det R = -1$, but we are not interested in transformations that are not continuous.

By taking the differential of (10.8), we can define the infinitesimal 4-vector dX, joining two space-time events, as

$$dX = (cdt, dx, dy, dz).$$

The inner product of dX with itself is

$$dX \cdot dX = (cdt, dx, dy, dz) \cdot (cdt, dx, dy, dz)$$

or in matrix notation

$$(dX)^{\mathrm{T}}\eta(dX) = \begin{pmatrix} cdt & dx & dy & dz \end{pmatrix} \begin{pmatrix} 1 & 0 & 0 & 0 \\ 0 & -1 & 0 & 0 \\ 0 & 0 & -1 & 0 \\ 0 & 0 & 0 & -1 \end{pmatrix} \begin{pmatrix} cdt \\ dx \\ dy \\ dz \end{pmatrix}.$$

On multiplying out the right-hand side, we obtain

$$dX \cdot dX = c^2 dt^2 - dx^2 - dy^2 - dz^2.$$

On comparing with the infinitesimal invariant interval (10.2), we see that $dX \cdot dX$ is invariant. Indeed, this illustrates an important result: the *inner product of 4-vectors is Lorentz invariant* (see Exercise 10.4).

Accordingly, 4-vectors can be classified in the same way that the separation of events were classified in Section 10.1. Thus a 4-vector $V \in \mathbb{M}^4$ is called

$$\begin{aligned} \text{time-like if} \quad & V \cdot V > 0 \\ \text{null or light-like if} \quad & V \cdot V = 0 \\ \text{space-like if} \quad & V \cdot V < 0. \end{aligned} \tag{10.18}$$

These vectors can be represented on the light cone depicted in Figure 10.3. It is not possible to transform from one type of vector to another; for example, a time-like vector cannot be transformed to a null vector or space-like vector, etc. However, it is possible to transform any 4-vector to its canonical form. So, for some 4-vector with components $V = (V^0, V^1, V^2, V^3)$ in some inertial frame S, we can perform a spatial rotation such that the x-axis is parallel to a 3-vector with components (V^1, V^2, V^3), which means that V^2 and V^3 will vanish. Thus:

$$\begin{aligned} V \text{ is time-like, canonical form is} \quad & (\pm V^0, 0, 0, 0) \\ V \text{ is null, canonical form is} \quad & (V^0, V^0, 0, 0) \\ V \text{ is space-like, canonical form is} \quad & (0, V^0, 0, 0), \end{aligned}$$

where for the time-like and space-like components, the magnitude is

$$V^0 = \sqrt{|V \cdot V|} \geq 0.$$

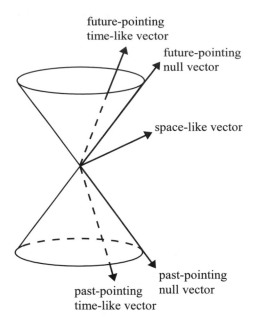

future-pointing
time-like vector

future-pointing
null vector

space-like vector

past-pointing
null vector

past-pointing
time-like vector

FIGURE 10.3
4-vectors with reference to the light cone

10.3.2   Four-velocity

Let us represent the world line of a particle in $\mathbb{M}^4$ by the trajectory

$$X = X(\tau) = (ct(\tau), \mathbf{x}(\tau))$$

which is the trajectory given in (10.9) but now parameterised by the proper time τ. Now, as $dX(\tau)$ is a time-like 4-vector and $d\tau$ is an invariant, we define another time-like 4-vector U called the *4-velocity* of the particle by

$$U = \frac{dX}{d\tau} = \left(c\frac{dt}{d\tau}, \frac{dx}{d\tau} \right). \tag{10.19}$$

With the aid of the chain rule, we can write

$$\frac{d\mathbf{x}}{d\tau} = \frac{d\mathbf{x}}{dt}\frac{dt}{d\tau},$$

where $d\mathbf{x}/dt$ is just the conventional 3-velocity of the particle, $\mathbf{u}$. Then, on incorporating (10.7) the 4-velocity of the particle becomes

$$U = \gamma(c, \mathbf{u}). \tag{10.20}$$

Note that for observers in S and S' whose coordinates are related by (10.10), their 4-velocities will be related by the transformation

$$U' = LU.$$

This is true because $d\tau$ is Lorentz invariant.

The inner product of U with itself can easily be formed using (10.20):

$$U \cdot U = \gamma^2(u)[c^2 - u^2]. \tag{10.21}$$

But the inner product of any two 4-vectors is Lorentz invariant, so (10.21) holds for all inertial frames. The 3-velocity of a particle in its rest frame must be zero. Thus on putting $u = 0$ ($\gamma = 1$)

$$U \cdot U = c^2. \tag{10.22}$$

As this must be true in all frames, the 4-velocity of any particle must conform to (10.22).

10.3.3 Four-acceleration

The 4-acceleration of a particle, A, is defined in terms of the 4-velocity as

$$A = \frac{dU}{d\tau}. \tag{10.23}$$

On using the chain rule, we have

$$A = \frac{dU}{dt}\frac{dt}{d\tau} = \gamma(u)\frac{dU}{d\tau}. \tag{10.24}$$

Differentiating (10.7) with respect to the coordinate t yields

$$\frac{dU}{dt} = \gamma(0, \mathbf{a}) + \dot{\gamma}(c, \mathbf{u}) = (\dot{\gamma}c, \gamma\mathbf{a} + \dot{\gamma}\mathbf{u}), \tag{10.25}$$

where $\dot{\gamma} \equiv d\gamma/dt$ and $\mathbf{a} \equiv d\mathbf{u}/dt$, the 3-acceleration of the particle. Substituting (10.25) into (10.24) gives

$$A = \gamma(\dot{\gamma}c, \gamma\mathbf{a} + \dot{\gamma}\mathbf{u}).$$

Because the particle is accelerating its frame will not be inertial. However, we may consider the particle to be instantaneously at rest, which would correspond to the particle's instantaneous rest frame; then its 3-velocity would be zero in this frame, $\mathbf{u} = 0$, and $\dot{\gamma}(0) = 0$. Consequently, the 4-acceleration becomes

$$A = (0, \mathbf{a}). \tag{10.26}$$

Although the 4-velocity U given by (10.20) will never vanish, regardless of which inertial frame we may be discussing, the 4-acceleration will vanish if and only if the 3-acceleration vanishes in the rest frame.

Notice that the 4-acceleration is a space-like 4-vector. Of interest is the inner product of the 4-velocity and the 4-acceleration. If we work in the rest frame ($\mathbf{u} = \mathbf{0}$) then, trivially,

$$U \cdot A = \gamma(c, \mathbf{0}) \cdot (0, \mathbf{a}) = 0,$$

(but see Exercise 10.5) which is the definition of orthogonality of 4-vectors.[4] So, the 4-velocity is always orthogonal to the 4-acceleration.

10.4   Exercises

1. The world line of a particle can be parameterised by the proper time τ such that $x = x(\tau)$, $t = t(\tau)$ in S and similarly in S'. Using the proper time, obtain the velocity transformation formula (9.31) for a particle performing non-uniform motion.

2. Defining a boost in the x-direction by

$$L(v) = \begin{pmatrix} \gamma & -\beta\gamma \\ -\beta\gamma & \gamma \end{pmatrix},$$

where $\beta = v/c$ and $\gamma = (1 - \beta^2)^{-1/2}$ show that the velocity transformation formula for successive boosts can be written as

$$L(v_1)L(v_2) = L(v_3),$$

where

$$v_3 = \frac{v_1 + v_2}{1 + \beta_1\beta_2}.$$

3. By writing a boost in the form

$$L(\rho) = \begin{pmatrix} \cosh\rho & -\sinh\rho \\ -\sin\rho & \cos\rho \end{pmatrix},$$

where ρ is the rapidity, show that the velocity transformation formula for successive boost can be written as

$$L(\rho_1)L(\rho_2) = L(\rho_3),$$

where $\rho_3 = \rho_1 + \rho_2$.

---

[4]Do not confuse *orthogonality* of 4-vectors with perpendicularity of 3-vectors.

4. For any two 4-vectors A and B in frame S use the Lorentz transformations to show that

$$A' \cdot B' = A \cdot B,$$

where A' and B' are 4-vectors in frame S'. This justifies the statement made in Section 10.3.1.

5. By differentiating (10.22) with respect to time show that the 4-velocity is orthogonal to the 4-acceleration, i.e.

$$U \cdot A = 0.$$

6. Prove that if two different null vectors are orthogonal they must also be parallel.

7. A tachyon (hypothetical faster-than-light particle)[5] has 4-velocity defined as $U = c(dX/ds)$, where $X = (ct, x, y, z)$ is the position 4-vector. Show that

$$U = \gamma(u)(c, \mathbf{u}),$$

where $\gamma(u) = 1/\sqrt{1 - (u^2/c^2)}$. Also, show that

$$U \cdot U = -c^2.$$

8. Show that a time-like vector can never be orthogonal to a null vector or to another time-like vector.

9. Let a particle in its rest frame S' have a 4-acceleration

$$A' = (0, \mathbf{a}'),$$

where $\mathbf{a}' = d\mathbf{u}'/dt'$ is the 3-acceleration and $\mathbf{u}'$ the 3-velocity. If the 3-velocity $\mathbf{u}$ and 3-acceleration $\mathbf{a}$ in S are parallel with the x-direction, use the Lorentz transformation $A = (L^{-1})A'$ to show that

$$(\gamma\dot{\gamma}c, \gamma(\dot{\gamma}u_x + \gamma a_x)) = (\gamma u_x a'_x/c, \gamma a'_x),$$

where the subscript x denotes quantities in the x-direction and $\gamma = \gamma(u_x)$. Further, show that the accelerations in S and S' are related by

$$a_x = \gamma^{-3} a'_x,$$

where $a_x = du_x/dt$.

---

[5] Actually, a tachyon indicates an instability of a theory in which it is present.

11

Relativistic Mechanics

CONTENTS

There is no question that the Galileo–Newton formulation of dynamics is inconsistent with the concepts of special relativity. Familiar dynamical quantities such as velocity and acceleration take on a whole new guise, Galilean invariance must now succumb to the more complicated Lorentz invariance, and underpinning everything discussed so far is the Einsteinian postulate that nothing can travel faster than the speed of light.[1]

We must now determine the Newtonian mechanical processes in terms of the language of special relativity.

11.1   Four-momentum

Crucial to our understanding of relativistic particle collisions is the notion of the 4-momentum, P. This is a multi-faceted quantity, and as we will presently see, it contains two very important relativistic quantities.

The 4-momentum is defined in terms of the 4-velocity, and is written as

$$P = mU = \gamma(u)(mc, m\mathbf{u}) \equiv (P^0, \mathbf{p}), \qquad (11.1)$$

---

[1] This statement should be qualified by saying that useful information cannot be transmitted faster than the speed of light. There are two common velocities that are associated with a wave: the group and phase velocities. The former transmits useful information and cannot, therefore, exceed the speed of light; the latter corresponds to the wave front and transmits no useful information, but can exceed the speed of light.

where m is called the *rest mass* (or just mass) of a particle. The spatial components of P are the components of the relativistic 3-momentum, $\mathbf{p}$:

$$\mathbf{p} = \gamma(u)m\mathbf{u}, \tag{11.2}$$

where $\mathbf{u}$ is the particle's 3-velocity.

The rest mass m is an invariant quantity in special relativity; that is, it remains invariant under Lorentz transformations. So the mass calculated in one inertial frame will have the same value in all inertial frames. Notice that as $\gamma \to 1$ (11.2) approaches the Newtonian limit.

[Note: It is not unusual (although there is little value in doing so) for the rest mass of a particle to be related to the so-called *relativistic* (or *inertial*) *mass* m_{rel}, by $m_{\mathrm{rel}} = \gamma m$. The reason that this is sometimes done is so that (11.2) may resemble the non-relativistic dynamical definition 'momentum = mass × velocity', and the equation that relates mass and energy can be written in its legendary form $E = m_{\mathrm{rel}}c^2$. However, attempting to contrive analogies between relativistic and non-relativistic equations, as in the former case, can be quite misleading. For example, relativistic kinetic energy cannot be written as $\frac{1}{2}m_{\mathrm{rel}}v^2$, nor force as $m_{\mathrm{rel}}\mathbf{a}$, unless the force happens to be orthogonal to the particle's 3-velocity.]

Equation (11.2) can be deduced from a gedankenexperiment involving ordinary (non-subatomic) particle collisions in two inertial frames S and S' say, moving relative to each other. It can be shown that conservation of Newtonian linear momentum holds in the frame of the fixed observer, but does not hold in the frame moving relative to the observer unless the Newtonian inertial mass is replaced by γm.

11.2 Relativistic energy

Although it is not obvious, the time component in (11.1), $P^0 \equiv \gamma mc$, will be found to be the energy divided by c of a freely moving particle with 4-momentum P. Therefore, let us define the relativistic energy as

$$E = P^0 c \equiv \gamma mc^2 \tag{11.3}$$

and show that it reduces to the classical kinetic energy for non-relativistic particles with velocities much less than the speed of light. By taking $u \ll c$, we can perform a Taylor expansion of γ:

$$\gamma = \left(1 - \frac{u^2}{c^2}\right)^{-1/2} = 1 + \frac{1}{2}\frac{u^2}{c^2} + \cdots .$$

Substituting this into (11.3) yields

$$E \approx mc^2 + \frac{1}{2}mu^2. \tag{11.4}$$

From a non-relativistic perspective (11.4) can be interpreted as the kinetic energy plus a constant mc^2. This has no effect upon the classical conservation of kinetic energy — it is merely a redefinition of the energy.

Example 11.1 Two particles P_1 and P_2 collide elastically (rest mass is preserved). Use conservation of relativistic energy and the energy definition (11.4) to show that the total kinetic energy is conserved in the system.

Solution Denote the masses of P_1 and P_2 before the collision as $\overline{m}_1$ and $\overline{m}_2$, respectively. Similarly, for the velocities of P_1 and P_2, we have u_1 and u_2 before, and $\overline{u}_1$ and $\overline{u}_2$ after. All barred quantities are assumed to be the values of these quantities after the collision.

Now, we can write conservation of relativistic energy as

$$E_1 + E_2 = \overline{E}_1 + \overline{E}_2. \tag{11.5}$$

Substituting (11.4) into this yields

$$m_1 c^2 + \frac{1}{2}m_1 u_1^2 + m_2 c^2 + \frac{1}{2}m_2 u_2^2 = \overline{m}_1 c^2 + \frac{1}{2}\overline{m}_1 \overline{u}_1^2 + \overline{m}_2 c^2 + \frac{1}{2}\overline{m}_2 \overline{u}_2^2$$

or

$$Mc^2 + T = \overline{M} c^2 + \overline{T},$$

where

$$M = m_1 + m_2, \quad T = \frac{1}{2}(m_1 u_1^2 + m_2 u_2^2).$$

As Newtonian mass is conserved, $M = \overline{M}$. Thence,

$$T = \overline{T},$$

which is the conservation of kinetic energy of the system. $\qquad\square$

Example 11.1 illustrates very well that in the classical limit conservation of relativistic energy (11.5), with relativistic energy defined by (11.3), reduces to the classical conservation of kinetic energy, which one would assume to be the case.

We can now rewrite (11.1) in a useful alternative form with the aid of (11.3):

$$P = (E/c, \mathbf{p}). \tag{11.6}$$

Thus in this form[2] we can see that conservation of 4-momentum yields conservation of relativistic energy and conservation of relativistic 3-momentum.

A very useful identity can be constructed from (11.1) and (11.6) by adhering to the invariant character of 4-vectors.

---

[2]Written like this P is sometimes called the *energy-momentum* 4-vector.

In a particle's rest frame the 3-velocity vanishes, $\mathbf{u} = \mathbf{0}$; consequently, the inner product formed from (11.1) gives

$$P \cdot P = (mc, \mathbf{0}) \cdot (mc, \mathbf{0})$$
$$= 7m^2c^2, \tag{11.7}$$

where, of course, $\gamma = 1$ in the rest frame. For all other inertial frames, the inner product formed from (11.6) gives

$$P \cdot P = \left(\frac{E}{c}, \mathbf{p}\right) \cdot \left(\frac{E}{c}, \mathbf{p}\right)$$
$$= \frac{E^2}{c^2} - \mathbf{p} \cdot \mathbf{p}$$
$$= \frac{E^2}{c^2} - p^2. \tag{11.8}$$

Because $P \cdot P$ is Lorentz invariant, it holds in any inertial frame, which means that we can equate (11.7) and (11.8) resulting in the identity

$$E^2 = p^2c^2 + m^2c^4. \tag{11.9}$$

11.3 Massless particles

So far our discussion of the 4-momentum has been restricted to massive particles (particles with mass). However, a mechanism is required so as to enable photons and other *massless particles*[3] to be incorporated within our current framework.

For a massless particle, we put $m = 0$ in (11.7) to give

$$P \cdot P = 0. \tag{11.10}$$

This implies that for massless particles the 4-momentum is null according to the criteria (10.18). More specifically, it is a future-pointing null 4-vector that lies only along the light cone.

Now, for a massless particle (11.9) can be written as

$$\frac{E}{c} = p. \tag{11.11}$$

---

[3]It was thought that all three types of neutrino (electron, tau and muon) were massless (this is the case according to the Standard Model). However, the absolute neutrino mass scale remains unknown, although the difference in the squares of their masses can be ascertained through experiment. The conclusion is that although feasible, it is unlikely that any neutrino is massless. Currently, the known massless particles are the photon and the graviton.

By making the prescription $\mathbf{p} = p\hat{\mathbf{n}}$, where $\hat{\mathbf{n}}$ is a unit vector specified in the direction of motion of the particle, the 4-momentum in (11.6) can be written as

$$P = \frac{E}{c}(1, \hat{\mathbf{n}}),\tag{11.12}$$

and this is the form most often employed in obtaining the energy and momentum of a massless particle.

Quantum mechanics brings to our attention the wave-like nature of all elementary particles, including the photon. That energy is not identical for all photons, but is related to the frequency of the electromagnetic wave associated with the photon was discovered by Max (Karl Ernst Ludwig) Planck (1858-1947) in 1900. The relation

$$E = h\nu,\tag{11.13}$$

where h is called *Planck's constant*, can be substituted into (11.12) giving an alternative form for the 4-momentum in terms of the frequency of the wave. It is, however, more usual to replace ν with the *angular frequency* $\omega = 2\pi\nu$, for which we can write

$$E = \hbar\omega,\tag{11.14}$$

where $\hbar = h/2\pi$. It was not until 1924 that Louis-Victor (Pierre Raymond, Duc De, Prinz) Broglie (1892-1987) postulated that elementary particles possessed a wave-like nature. He formulated a relation that allowed the particle's linear momentum to be given in terms of the wavelength of wave associated with the particle:

$$p = \frac{h}{\lambda}.\tag{11.15}$$

With the above relations, the 4-momentum of the photon can be written as

$$P = \frac{h\nu}{c}(1, \hat{\mathbf{n}}) = \frac{\hbar\omega}{c}(1, \hat{\mathbf{n}}) = \frac{h}{\lambda}(1, \hat{\mathbf{n}}).\tag{11.16}$$

The usefulness of these relations will be seen in the following sections.

11.4 Aberration

The phenomenon known as *aberration* was observed for the first time by the English astronomer James Bradley (1693-1762) in 1728. Bradley noticed whilst conducting experiments to observe parallax in the position of a distant star, that the change in position of the star was greatest between the months of September and March. However, this contrasted with the expected result from parallax, which gave the greatest change between December and June. This effect is due to the change in the angle of incidence of the light emitted by a distant star due to the motion of the Earth, and is called aberration.

Of course, aberration can have huge consequences for observational astronomy, and as photons are relativistic particles, a full relativistic treatment of this phenomenon is desirable.

11.4.1   The Lorentz transformations for four-momentum

As with all 4-vectors, 4-momentum must transform according to the Lorentz transformations:

$$P' = LP \tag{11.17}$$

or, explicitly in terms of relativistic energy and 3-momentum,

$$\begin{pmatrix} E'/c \\ p'_x \\ p'_y \\ p'_z \end{pmatrix} = \begin{pmatrix} \gamma & -\beta\gamma & 0 & 0 \\ -\beta\gamma & \gamma & 0 & 0 \\ 0 & 0 & 1 & 0 \\ 0 & 0 & 0 & 1 \end{pmatrix} \begin{pmatrix} E/c \\ p_x \\ p_y \\ p_z \end{pmatrix}. \tag{11.18}$$

Consider two inertial frames S and S' with S' moving at velocity v relative to S. A light source at rest in S' is positioned at O'. At $t = t' = 0$ the source emits a photon at an angle θ in S and θ' in S' with respective energies E and E' as depicted in Figure 11.1. In the diagram, $\hat{\mathbf{n}}$ and $\hat{\mathbf{n}}'$ are the respective directions of motion of the photon in S and S'.

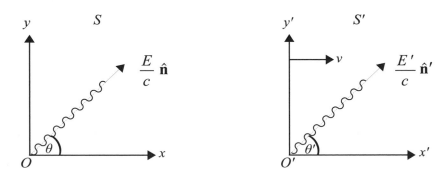

FIGURE 11.1
Photon emitted by a source at rest in S' as seen in S and S'

With the photon depicted in this configuration, the Lorentz transformations (11.12) can be written as

$$\begin{pmatrix} E'/c \\ (E'/c)\cos\theta' \\ (E'/c)\sin\theta' \\ 0 \end{pmatrix} = \begin{pmatrix} \gamma & -\beta\gamma & 0 & 0 \\ -\beta\gamma & \gamma & 0 & 0 \\ 0 & 0 & 1 & 0 \\ 0 & 0 & 0 & 1 \end{pmatrix} \begin{pmatrix} E/c \\ (E/c)\cos\theta \\ (E/c)\sin\theta \\ 0 \end{pmatrix}.$$

On multiplying out the matrix components, we obtain the following relativistic energy transformations:

$$E' = \gamma E(1 - \beta \cos \theta) \tag{11.19}$$
$$E' \cos \theta' = -\gamma E(\beta - \cos \theta) \tag{11.20}$$
$$E' \sin \theta' = E \sin \theta. \tag{11.21}$$

In fact, (11.19) is the only transformation required to enable us to calculate the energy of the photon between frames. Note, on dividing this by c, we obtain the photon momentum transformation relation. Alternatively, on replacing the energy with (11.13), a relation for the photon frequency is obtained:

$$\nu' = \gamma \nu (1 - \beta \cos \theta), \tag{11.22}$$

and, similarly, for the wavelength of the photon,

$$\frac{1}{\lambda'} = \frac{\gamma}{\lambda}(1 - \beta \cos \theta). \tag{11.23}$$

Notice that h does not feature in any of these relations on account of it being invariant under a Lorentz transformation.

A relation for the angle at which the photon is emitted can be established by combining (11.19) and (11.20). Thus,

$$\cos \theta' = \frac{\cos \theta - \beta}{1 - \beta \cos \theta} \tag{11.24}$$

or, by using the trigonometric identity

$$\cos \alpha = \frac{1 - \tan^2(\alpha/2)}{1 + \tan^2(\alpha/2)},$$

we have

$$\tan \frac{\theta'}{2} = \tan \frac{\theta}{2}\sqrt{\frac{1 + \beta}{1 - \beta}}. \tag{11.25}$$

This is the aberration formula and it gives the observed change in direction of a light beam in two different inertial frames. Note that if the light beam were incoming (towards the origin) $\beta \to -\beta$ in the above formulae. In the above analysis, we considered only the motion of a single particle, a photon. However, if one were to observe the uniform motion of an extended body, the light beams being reflected off the body would be deflected by the eye at different instances of time; thus aberration causes the object to be visually distorted.

11.4.2   The relativistic Doppler effect

Most of us have experienced the *Doppler effect*[4] at some time or another. For example, when standing on a station platform the pitch of a train's whistle drops suddenly as the train rushes past us. Thus the detectable change in frequency of a sound source resulting from the relative motion of the source and the observer is known as the Doppler effect. This is also true for a light source. Although not explicitly stated, (11.22) is the Doppler effect for photons emitted in a general direction.

Two cases arise when we consider a light source moving radially and transversely with respect to a stationary observer.

For the *radial* case, we consider a light source at rest in S' that emits a light beam at an angle $\theta' = 0$ (refer to Figure 11.1). An observer at rest in S will detect the approaching light to have a frequency ν_{approach}, where

$$\nu_{\text{approach}} = \frac{\nu_0}{\gamma(1 - \beta)} = \nu_0 \sqrt{\frac{1 + \beta}{1 - \beta}}. \tag{11.26}$$

Note that ν' is written as ν_0 to indicate the *proper frequency* of the light source.

For the source emitting a light beam at an angle $\theta' = \pi$, an observer at rest in S will detect the receding light to have a frequency ν_{recede}, where

$$\nu_{\text{recede}} = \frac{\nu_0}{\gamma(1 + \beta)} = \nu_0 \sqrt{\frac{1 - \beta}{1 + \beta}}. \tag{11.27}$$

If the velocity of the source is much less than the speed of light, (11.26) and (11.27) reduce to the standard non-relativistic Doppler effect for a moving source:

$$\nu_{\text{approach}} = \frac{\nu_0}{1 - \beta}, \quad \nu_{\text{recede}} = \frac{\nu_0}{1 + \beta}.$$

For the *transverse* case, we consider a light source at rest in S' that emits a light beam at an angle $\theta' = \pi/2$. An observer at rest in S will detect the light beam crossing their frame at a right angle, and so no radial component will be detected. The frequency of the light $\nu_{\text{transverse}}$ detected by the observer is given by

$$\nu_{\text{transverse}} = \frac{\nu_0}{\gamma} = \nu_0 \sqrt{1 - \beta^2}. \tag{11.28}$$

Notice that on putting $\theta = \pi/2$ in (11.22), $\nu' = \gamma\nu$. Because the γ factor is purely relativistic, there is no transverse Doppler effect in non-relativistic physics. In fact, the transverse Doppler effect has been utilised successfully in experiments to detect time dilation.

---

[4]Christian Johann Doppler (1803-1853) was an Austrian physicist whose main contribution to physics was the discovery of the effect named in his honour.

11.5 Particle collisions

Special relativity really comes into its own when one places it within the context of particle physics. Indeed, as one would expect, special relativity is an essential ingredient in order to understand the kinematical behaviour of elementary particle collisions that take place within the immense particle accelerator rings.

11.5.1 Compton scattering

In 1923, the American physicist Arthur Holly Compton (1892-1962) discovered that the wavelength of X-rays increases as they pass through paraffin wax. He concluded that this increase was brought about because of photons being scattered by stationary electrons during this elastic collision process. This important result confirmed Einstein's belief that photons possessed energy and momentum.

The collision process can be expressed rather simply by

$$\gamma + e^- \to \gamma + e^-,$$

where γ is the symbol for the photon (not to be confused with the same symbol for the relativistic factor) and e^- is the symbol for the electron. The process whereby photons are scattered by stationary electrons is called *Compton scattering* and is depicted in Figure 11.2 in the LAB frame.

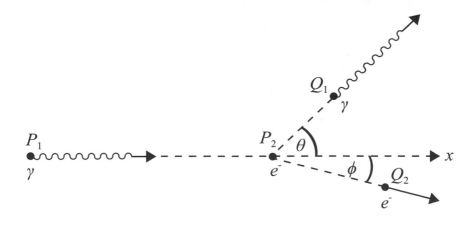

FIGURE 11.2
Compton scattering in the LAB frame

The photon with 4-momentum P_1 collides with an electron at rest, with

4-momentum P_2. After the collision the photon scatters at a scattering angle θ_1 and 4-momentum Q_1, while the electron recoils at a recoil angle θ_2 and 4-momentum Q_2.

Example 11.2 With reference to Figure 11.2, use conservation of 4-momentum and the invariance property of 4-vectors to obtain the formula

$$\lambda_2 - \lambda_1 = \lambda_c(1 - \cos\theta), \qquad (11.29)$$

where λ_1 is the wavelength of the photon before scattering, λ_2 is the wavelength of the photon after scattering and the *Compton wavelength* is given by $\lambda_c = h/m_e c$, with m_e being the rest mass of the electron.

Solution From the definition (11.16), we can write the 4-momentum of the photon before and after the collision as

$$P_1 = \frac{h}{\lambda_1}(1,1,0,0)$$

and

$$Q_1 = \frac{h}{\lambda_2}(1,\cos\theta,\sin\theta,0),$$

where the initial 4-momentum of the photon is along the x-axis. From the definition (11.1), we can write the 4-momentum of the electron before and after the collision as

$$P_2 = m_e(c,\mathbf{0})$$

(note $\gamma(|\mathbf{u}|) = 1$ because $\mathbf{u} = \mathbf{0}$) and

$$Q_2 = \gamma m_e(c,\mathbf{u_e}),$$

where $\mathbf{u_e}$ is the 3-velocity of the electron after the collision.

By conservation of 4-momentum

$$P_1 + P_2 = Q_1 + Q_2$$

or, since we are not interested in Q_2,

$$(P_1 + P_2 - Q_1)^2 = Q_2^2.$$

On expanding, we get

$$P_1^2 + P_2^2 + Q_1^2 + 2P_1 \cdot P_2 - 2P_1 \cdot Q_1 - 2P_2 \cdot Q_1 = Q_2^2.$$

Because P_1 and Q_1 are null vectors, we have $P_1^2 = 0 = Q_1^2$. Also, $P_2^2 = Q_2^2$ (see Exercise 11.2). The expression above then reduces to

$$P_1 \cdot P_2 - P_1 \cdot Q_1 - P_2 \cdot Q_1,$$

and on substituting in the appropriate expression for the 4-momenta, the result follows. □

11.5.2 Collisions in the LAB frame and CM frame

It was shown in Section 6.5 that it is useful to have representations of particle collisions in both the LAB frame and CM frame. Similarly, we will follow this pattern here and describe a particle collision in the LAB frame, which will be denoted by S_{LAB}, and the CM frame, which will be denoted by S'_{CM}. Both frames are inertial.

In S_{LAB}, particle 1 has rest mass m and moves along the x-axis with 4-momentum P_1. This particle collides with particle 2 with 4-momentum P_2. Note, as usual, it will be assumed that particle 2 is at rest in S_{LAB} so that its 3-velocity $\mathbf{u}_2$ and, therefore, 3-momentum $\mathbf{p}_2$ are zero. After the collision, particle 1 moves with 4-momentum P_3 at an angle θ, while particle 2 moves with 4-momentum P_4 at an angle ϕ. Figure 11.3 depicts the collision in S_{LAB}. The various 4-momenta are written as

$$P_1 = \left(\frac{E_1}{c}, \mathbf{p}_1\right), \quad P_2 = \left(\frac{E_2}{c}, \mathbf{0}\right), \quad P_3 = \left(\frac{E_3}{c}, \mathbf{p}_3\right), \quad P_4 = \left(\frac{E_4}{c}, \mathbf{p}_4\right),$$

$$(11.30)$$

where the 3-momentum $\mathbf{p}_1 = (p_{1x}, p_{1y}, p_{1z})$, and similarly for the other 3-momenta.

In S'_{CM} (the frame in which the total 3-momentum is zero), particle 1 moves along the x'-axis (which coincides with the x-axis) with 4-momentum P'_1 and collides with particle 2 moving in the opposite direction with 4-momentum P'_2. The particles must be travelling towards each other with equal speed. Let the 3-velocity of particle 1 be $\mathbf{u}'_1 = (v, 0, 0)$ then S'_{CM} is moving at velocity v relative to S_{LAB}. After the collision, particle 1 moves with 4-momentum P'_3 at an angle θ', while particle 2 moves with 4-momentum P'_4 at the same angle (relative to the x-axis) but in the opposite direction. Figure 11.3 depicts the collision in S'_{CM}. The various 4-momenta are written as

$$P'_1 = \left(\frac{E'_1}{c}, \mathbf{p}'_1\right), \quad P'_2 = \left(\frac{E'_2}{c}, \mathbf{p}'_2\right), \quad P'_3 = \left(\frac{E'_3}{c}, \mathbf{p}'_3\right), \quad P'_4 = \left(\frac{E'_4}{c}, \mathbf{p}'_4\right).$$

$$(11.31)$$

Now, the total relativistic energy in S_{LAB} before the collision is, by (11.30),

$$E = E_1 + E_2 = \gamma(u_{1x})mc^2 + mc^2 = (\gamma(u_{1x}) + 1)mc^2; \quad (11.32)$$

note: $\mathbf{u}_2 = 0$, therefore $\gamma(\mathbf{u}_{2x}) = 1$. The total 3-momentum in S_{LAB} before the collision is

$$\mathbf{p} = \mathbf{p}_1 + \mathbf{p}_2 = \mathbf{p}_1 = (p_{1x}, 0, 0) = (\gamma(u_{1x})mu_{1x}, 0, 0). \quad (11.33)$$

From the transformation (11.18), the 3-momentum components transform between S_{LAB} and S'_{CM} according to

$$p'_x = \gamma\left(p_x - \frac{\beta E}{c}\right), \quad p'_y = p_y, \quad p'_z = p_z.$$

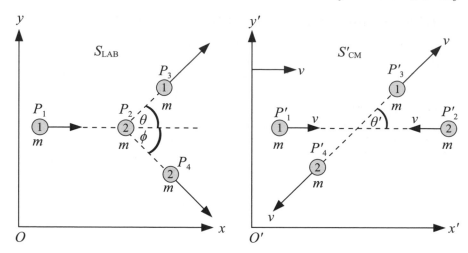

FIGURE 11.3
Collision of a pair of particles in S_{LAB} and S'_{CM}

As the total 3-momentum must be zero in S'_{CM}, we have

$$p_{1x} = \frac{\beta(v)E}{c} = \beta(v)[\gamma(u_{1x}) + 1]mc, \quad p_{1y} = 0, \quad p_{1z} = 0.$$

On comparing the p_{1x} component with (11.33), we have

$$\gamma(u_{1x})mu_{1x} = \beta(v)[\gamma(u_{1x}) + 1]mc, \tag{11.34}$$

or, in terms of v,

$$v = \frac{\gamma(u_{1x})u_{1x}}{\gamma(u_{1x}) + 1}.$$

This is the velocity of S'_{CM} relative to S_{LAB} in terms of the initial velocity of particle 1.

The total relativistic energy in S'_{CM} before the collision is, by (11.31),

$$E' = E'_1 + E'_2 = 2\gamma(v)mc^2.$$

As 3-momentum is conserved in this frame and relativistic energy is also conserved:

$$E'_1 = E'_2 = E'_3 = E'_4 = \gamma(v)mc^2.$$

Motion takes place only in the $x'y'$-plane. Thus the components of $\mathbf{u}'_3$ are

$$\mathbf{u}'_3 = (u'_{3x}, u'_{3y}, u'_{3z}) = (v \cos\theta', v \sin\theta', 0).$$

Using the velocity transformation formulae (9.29), we have, in S_{LAB},

$$u_{3x} = \frac{v(\cos\theta' + 1)}{1 + \beta^2 \cos\theta'}, \quad u_{3y} = \frac{v \sin\theta'}{\gamma(1 + \beta^2 \cos\theta')}, \quad u_{3z} = 0.$$

But
$$\mathbf{u}_3 = (u_{3x}, u_{3y}, u_{3z}) = (|\mathbf{u}_3| \cos \theta, |\mathbf{u}_3| \sin \theta, 0).$$

Hence,
$$\tan \theta = \frac{u_{3y}}{u_{3x}} = \frac{\sin \theta'}{\gamma(1 + \cos \theta')}. \tag{11.35}$$

The components of $\mathbf{u}_4'$ are
$$\mathbf{u}_4' = (u_{4x}', u_{4y}', u_{4z}') = (-v \cos \theta', -v \sin \theta', 0).$$

Using again (9.29)
$$u_{4x} = \frac{v(1 - \cos \theta')}{1 - \beta^2 \cos \theta'}, \quad u_{4y} = \frac{-v \sin \theta'}{\gamma(1 - \beta^2 \cos \theta')}, \quad u_{4z} = 0.$$

But
$$\mathbf{u}_4 = (u_{4x}, u_{4y}, u_{4z}) = (|\mathbf{u}_4| \cos \phi, -|\mathbf{u}_4| \sin \phi, 0).$$

Hence,
$$\tan \phi = -\frac{u_{4y}}{u_{4x}} = \frac{\sin \theta'}{\gamma(1 - \cos \theta')}. \tag{11.36}$$

On multiplying (11.35) by (11.36), we have
$$\tan \theta \tan \phi = 1 - \beta^2(v).$$

But from (11.34),
$$\beta(v) = \frac{\gamma(u_{1x})\beta(u_{1x})}{1 + \gamma(u_{1x})}.$$

Thus,
$$\tan \theta \tan \phi = \frac{2}{1 + \gamma(u_{1x})}.$$

In the classical Newtonian limit $u_{1x} \ll c$, $\tan \theta \tan \phi \approx 1$. This implies that
$$\cos \theta \cos \phi - \sin \theta \sin \phi \approx 0$$

or
$$\cos(\theta + \phi) \approx 0$$

giving
$$\theta + \phi = \frac{\pi}{2},$$

which is the result, (6.63), established for non-relativistic collisions. Note that the above analysis cannot be conducted for any pair of elementary particles. Indeed, quantum theory forbids this kind of scattering if the particles are electrons.

Example 11.3 A particle with 4-momentum $P_1 = (E_1/c, p_1, 0, 0)$,

p_1 being the x component of the 3-momentum, collides with a photon head-on along the x-axis with energy E_2. Show that the recoil energy of the photon is

$$\frac{E_2(E_1 + p_1 c)}{2E_2 + E_1 - p_1 c}.$$

Solution Let P_1 and P_2 be the 4-momentum of the particle and photon before the collision, and let P_3 and P_4 be the 4-momentum of the particle and photon after the collision. Thus, the components of these 4-momenta are

$$P_1 = \left(\frac{E_1}{c}, p_1, 0, 0\right), \quad P_2 = \left(\frac{E_2}{c}, -\frac{E_2}{c}, 0, 0\right),$$

$$P_3 = \left(\frac{E_3}{c}, p_3, 0, 0\right), \quad P_4 = \left(\frac{E_4}{c}, -\frac{E_4}{c}, 0, 0\right).$$

By conservation of 4-momentum

$$P_1 + P_2 = P_3 + P_4$$

or, since we are not interested in P_3, on rearranging and squaring

$$P_1^2 + P_2^2 + P_4^2 + 2P_1 \cdot P_2 - 2P_1 \cdot P_4 - 2P_2 \cdot P_4 = P_3^2.$$

This can be reduced if we note the following:

$$P_1^2 = P_3^2 = mc^2 \quad m \text{ being the particle's mass,}$$
$$P_2^2 = P_4^2 = 0 \quad\quad P_2 \text{ and } P_4 \text{ being null.}$$

Therefore,

$$P_1 \cdot P_2 = (P_1 - P_2) \cdot P_4.$$

On substituting in the appropriate 4-momenta, we have

$$E_2(E_1 + p_1 c) = (E_1 - p_1 c + 2E_2)E_4,$$

giving

$$E_4 = \frac{E_2(E_1 + p_1 c)}{2E_2 + E_1 - p_1 c}. \quad \square$$

11.5.3   Threshold energy

In most instances elementary particle collisions have resulted in the production of other sometimes new, elementary particles. For example, the proton-proton collision

$$p + p \rightarrow p + p + (p + \bar{p})$$

resulted in the production of another proton p plus an antiproton $\bar{p}$.[5] Reactions such as these require the particles before the collision to have a minimum energy; this is called the *threshold energy*.

Let the 4-momentum for each proton and antiproton be P_i, $i = 1, 2, 3, 4, 5, 6$. Then from the conservation of 4-momentum

$$P_1 + P_2 = P_3 + P_4 + P_5 + P_6.$$

On squaring both sides

$$(P_1 + P_2)^2 = (P_3 + P_4 + P_5 + P_6)^2.$$

We will consider the collision between the protons with 4-momentum P_1 and P_2, first in S'_{CM} and then in S_{LAB}.

In the S'_{CM} frame, we have the total 3-momentum zero: the protons will approach each other with $|\mathbf{p}|$. Thus taking

$$P_1 = \left(\frac{E'}{c}, \mathbf{p}'\right) \text{ and } P_2 = \left(\frac{E'}{c}, -\mathbf{p}'\right),$$

we have

$$(P_1 + P_2)^2 = 4\frac{E'^2}{c^2}. \tag{11.37}$$

Now, the total energy of the system after the collision must be

$$E' = \sum_{i=1}^{4} \gamma(u_i) m_i c^2,$$

where u_i is the velocity of each particle and $m_1 = m_2 = m_3 = m_4 = m$. (Note that the rest mass of the antiproton is the same as that of the proton.) The range of i corresponds to the number of particles after the collision. To eventually obtain the threshold energy, the minimum value of E' is required after the collision. This occurs when the 3-momentum of each particle is zero; consequently, each γ must take the value of unity. Hence

$$E'_{\text{min}} = 4mc^2. \tag{11.38}$$

In the S_{LAB} frame, we take

$$P_1 = \left(\frac{E}{c}, \mathbf{p}\right) \text{ and } P_2 = (mc, \mathbf{0}).$$

Remember that this is the frame in which one of the colliding particles is at rest. Thence,

$$(P_1 + P_2)^2 = P_1^2 + P_2^2 + 2P_1 \cdot P_2$$

$$= m^2 c^2 + m^2 c^2 + 2mc\left(\frac{E}{c}\right)$$

$$= 2m^2 c^2 + 2mE. \tag{11.39}$$

---

[5]This reaction was the first to produce an antiparticle of any kind.

Note that we are employing the fact that for any massive particle of mass m the square of its corresponding momentum is m^2c^2 (see Exercise 11.2).

Because, of course, $(P_1 + P_2)^2$ transforms invariantly between inertial frames, we can equate (11.37) and (11.39):

$$4\frac{E'^2}{c^2} = 2m^2c^2 + 2mE, \tag{11.40}$$

where E is the total proton energy before the collision. For this to be the threshold energy, we must now replace E' with $E'_{\min}$ given in (11.38). On doing this and rearranging (11.40), we get

$$E_{\text{threshold}} = 7mc^2.$$

The minimum kinetic energy required to produce antiprotons from the reaction

$$p + p \to p + p + (p + \bar{p})$$

is given by

$$T = E_{\text{threshold}} - mc^2 = 6mc^2.$$

The value of this is approximately 5600 MeV. A problem arises when the LAB frame is used to produce high energy particles. Performing collisions between electrons and positrons, say, where one or the other is at rest to produce particles with mass greater than $(3000/c^2)$ MeV, would require threshold energy of the order of 10^7 MeV. This is way beyond the scope of existing particle accelerators. However, by conducting the collision in the CM frame, both particles are now accelerated towards each other; the result is that a significantly lower threshold energy is required to produce particles with a very large mass.

11.6 Exercises

1. Starting with the definition of relativistic energy (11.3), obtain the identity (11.9):
$$E^2 = p^2c^2 + m^2c^4.$$

2. Show that $P_2^2 = Q_2^2 = m_e^2c^2$ in Example 11.2.

3. With reference to Figure 11.2, use conservation of 4-momentum (without the aid of the invariance property of 4-vectors) to obtain formula (11.29). [Hint: You will need to consider motion parallel and perpendicular to the x-axis.]

4. Two protons 1 and 2 of mass m collide elastically as observed in S_{LAB} and then in S'_{CM}. Figure 11.4 depicts the collision in each frame. Let the energy of the proton in S'_{CM} be E' and assume that

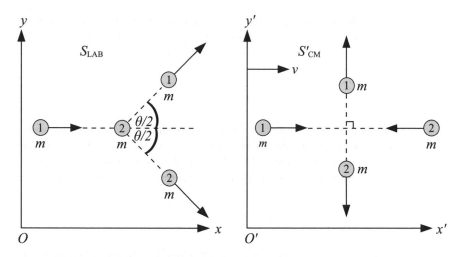

FIGURE 11.4
Proton-proton collision in S_{LAB} and S'_{CM}

S'_{CM} is moving relative to S_{LAB} along the coincident axes x and x' with velocity v. Show that

$$\frac{1}{\gamma(v)} = \tan\left(\frac{\theta}{2}\right).$$

Also, show that in S_{LAB} proton 1 has energy

$$E = \frac{2E'^2 - m^2 c^4}{mc^2}$$

and that

$$\tan\left(\frac{\theta}{2}\right) = \frac{E'}{mc^2}.$$

5. A photon collides with and is absorbed by a proton at rest in the LAB frame S_{LAB}. The collision produces a neutral pion and a proton:

$$\gamma + p \rightarrow p + \pi^\circ.$$

Using both S_{LAB} and the CM frame S'_{CM}, show that the threshold energy of the reaction is given by

$$E_{\text{threshold}} = \left(\frac{m_{\pi^\circ}^2}{2m_{\text{p}}} + 1\right)c^2,$$

where m_{π° is the mass of the pion and m_{p} is the mass of the proton.

6. A particle of rest mass m_1 moving with velocity u collides with a particle m_2 in the rest frame of m_2. On colliding, the particles coalesce. Show that the speed of the composite particle is

$$\frac{\gamma(u)m_1 u}{\gamma(u)m_1 + m_2}.$$

7. A rocket is designed so that it can travel in a straight line by emitting photons in one direction along the line, thus propelling it in the opposite direction. Show that the initial rest mass of the rocket m_i and the final rest mass of the rocket m_f are related by the equation

$$m_i = m_f \left(\frac{c+v}{c-v} \right)^{1/2},$$

where v is the velocity of the rocket when the mass of the rocket is m_f. [Note: Use conservation of 4-momentum.]

A

Appendix: Conic Sections

CONTENTS

The circle, ellipse, hyperbola and parabola encountered in Chapter 5, whilst discussing orbits governed by an inverse square law of force, are all examples of *conic sections*, or *conics* for short. They arise when a right-circular cone is intersected in a particular way by a plane. Figure A.1 shows the different configurations of the plane intersecting the cone.

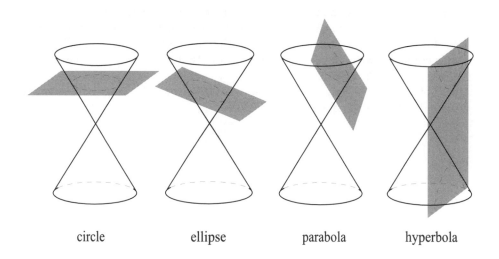

     circle   ellipse   parabola   hyperbola

FIGURE A.1
Conic sections

A.1   Conic sections in Cartesian coordinates

The cone and the plane can be represented in Cartesian coordinates by a second-order and first-order equation, respectively. Thus the most general form of a conic section is given by the second-order equation

$$Ax^2 + Bxy + Cy^2 + Dx + Ey + F = 0,$$

where $A, B, \ldots, F$ are constants, and $A^2 + B^2 + C^2 > 0$. However, a little care must be observed due to the fact that if the left-hand side of the equation can be factorised as

$$(ax + by + c)(dx + ey + f) = 0,$$

where $a, b, \ldots, f$ are constants, the equation represents a pair of straight lines, a single straight line, a point or the empty set. The conic sections themselves, however, can be classified according to the values of the real constants $A, B, \ldots, F$; although in some cases a lengthy calculation is required that involves a method of matrix manipulation known as diagonalisation,[1] which is beyond the scope of this book.

A.1.1   Parabolas

A *parabola* is defined as the locus of points in a plane that are equidistant from a fixed point F, known as the *locus*, and a fixed line D, known as the *directrix*. It is an open curve that is symmetrical about the line, known as the *axis*. The point at which the parabola cuts the axis is known as the *vertex*.

Let us orientate the parabola such that its vertex is at the origin and its axis coincides with the x-axis, which means that the directrix is parallel with the y-axis (see Figure A.2).

Let $P(x, y)$ be any point on the parabola and $Q(-a, y)$ be the nearest point from P on D. Thus, by definition,

$$|FP| = |PQ|$$

or

$$\sqrt{(x - a)^2 + y^2} = x + a.$$

Squaring both sides and simplifying yields

$$y^2 = 4ax,$$

which is the equation of the parabola.

The length of the chord through F and parallel with D is the latus rectum, which has the value of $4a$.

---

[1]The reader should consult any of the modern texts on linear algebra for a description of the diagonalisation process.

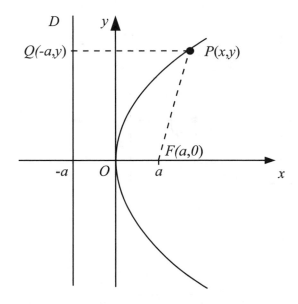

FIGURE A.2
The parabola: $y^2 = 4ax$

Notice that on rotating Figure A.2 clockwise successively about an axis through the origin by $\pi/2$, keeping the axes fixed, we obtain the following equations of a parabola:

$$x^2 = -4ay, \text{ on rotating by } \pi/2$$
$$y^2 = -4ax, \text{ on rotating by } \pi$$
$$x^2 = 4ay, \text{ on rotating by } 3\pi/2.$$

A.1.2 Ellipses

An *ellipse* is defined as the locus of points in a plane, where the sum of the distances from two fixed points F_1 and F_2, known as *foci*, to any point on the line is a constant.

Let us orientate an ellipse such that the foci lie on the x-axis at the points $F_1(-c, 0)$ and $F_2(c, 0)$. The origin is then equidistant to F_1 and F_2 (see Figure A.3).

Let $P(x, y)$ be any point on the ellipse and the sum of the distances from P to the foci be $2a$, $a > 0$. Thus, by definition,

$$|F_2P| + |F_1P| = 2a$$

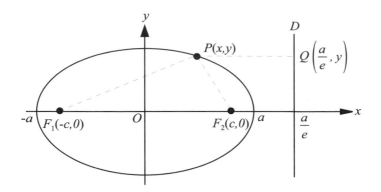

FIGURE A.3
The ellipse: $x^2/a^2 + y^2/b^2 = 1$, $a > b$

or
$$\sqrt{(x-c)^2 + y^2} + \sqrt{(x+c)^2 + y^2} = 2a$$

or
$$(x-c)^2 + y^2 = (2a - \sqrt{(x+c)^2 + y^2})^2.$$

Expanding the right-hand side and simplifying yields
$$a\sqrt{(x+c)^2 + y^2} = a^2 + cx$$

or
$$(a^2 - c^2)x^2 + a^2 y^2 = a^2(a^2 - c^2).$$

Finally, on letting $b^2 = a^2 - c^2$ and rearranging, we get
$$\frac{x^2}{a^2} + \frac{y^2}{b^2} = 1, \quad a > b,$$

which is the equation of the ellipse.

Any chord that passes through the origin is a diameter. The ellipse has two diameters that act as axes of symmetry: the *major axis*(being the largest diameter), and the *minor axis* (being the shortest diameter). The line segment joining the origin to the ellipse along the major axis is called the *semi-major axis*, a; similarly, the *semi-minor axis* is b. The two points at which the major axis intersects the ellipse are known as *vertices*. The vertices can be easily found by putting $y = 0$ in the equation of the ellipse, yielding $x = \pm a$. Similarly for $x = 0$, $y = \pm b$.

For the case $a = b$, we then have $c = 0$ and the foci coincide. Thus the ellipse degenerates to a circle:
$$x^2 + y^2 = a^2, \quad a = b \neq 0.$$

Each focus of the ellipse is a distance ae from O, where e is known as the eccentricity e. For the ellipse in Figure A.3 the directrix D is a distance a/e from the origin, where $e = c/a$. In general, the eccentricity is given by the ratio F_2P/PQ.[2] As Q lies on the directrix D, PQ is perpendicular to D and, therefore, parallel to the x-axis. So

$$|F_2P|^2 = (x - c)^2 + y^2$$
$$= x^2 - 2cx + c^2 + b^2(1 - \frac{x^2}{a^2})$$
$$= \frac{c^2 x^2}{a^2} - 2cx + a^2$$
$$= (a - c/ax)^2$$
$$= (a - ex)^2$$
$$\therefore F_2P = a - ex.$$

From Figure A.3, we have

$$PQ = \frac{a}{e} - x = \frac{(a - ex)}{e}.$$

Hence,

$$\frac{F_2P}{PQ} = e.$$

Note that for any ellipse $O \le e < 1$, and $e = 0$ gives a circle. Either of the chords through F_1 or F_2 and parallel to the directrix is a latus rectum, the length being $2b^2/a$.

A.1.3 Hyperbolas

A *hyperbola* is defined as the locus of points in a plane for which the difference of the distances from two fixed points F_1 and F_2 (the foci) is constant.

Let us orientate a hyperbola such that the foci lie on the x-axis at the points $F_1(-c, 0)$ and $F_2(-c, 0)$. The origin is then equidistant to F_1 and F_2 (see Figure A.4).

If the foci of the hyperbola are situated at the points $F_1(-c, 0)$ and $F_2(-c, 0)$ and the difference from $P(x, y)$ to the foci is $2a$, $c > a$, then

$$|F_1P| - |F_2P| = \pm 2a$$

or

$$\sqrt{(x + c)^2 + y^2} - \sqrt{(x - c)^2 + y} = \begin{cases} 2a & \text{for the right branch} \\ -2a & \text{for the left branch.} \end{cases}$$

---

[2]If the directrix had been on the other side of the y-axis, we would have $e = F_1P/PQ$ for the point $Q(-u/e, y)$.

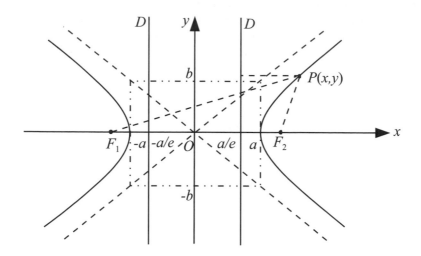

FIGURE A.4
The hyperbola: $x^2/a^2 - y^2/b^2 = 1$

After a little algebra, we have

$$\frac{x^2}{a^2} - \frac{y^2}{b^2} = 1, \qquad b^2 = c^2 - a^2.$$

The vertices lie on the hyperbola at $x = \pm a$, where a is the *semi-transverse axis* (which we called the semi-major axis in Section 5.5). The term *semi-conjugate axis* is used for b. The two dotted diagonal lines that pass through the rectangle, which is centred at the origin, are asymptotes of the hyperbola and are solutions of the equation

$$\frac{x^2}{a^2} - \frac{y^2}{b^2} = 0.$$

Thus the asymptotes of the hyperbola are the lines

$$y = \frac{b}{a}x \quad \text{and} \quad y = -\frac{b}{a}x.$$

If $a = b$, the equation of the hyperbola becomes

$$x^2 - y^2 = a^2.$$

The asymptotes are then mutually perpendicular and are the lines

$$y = x \quad \text{and} \quad y = -x.$$

A hyperbola as such is referred to as a *rectangular hyperbola*.

Each focus of the hyperbola is a distance ae from the origin. For the hyperbola in Figure A.4, the directrices, each labelled D, are a distance a/e from the origin, where $e = c/a$. If we concentrate on the right branch of the hyperbola, the eccentricity is given by the ratio F_2P/PQ. A similar calculation to that which was close for the ellipse can easily prove this assertion, and is left as an exercise for the reader. Note that for any hyperbola $e > 1$. Either of the two chords through F_1 and F_2 and parallel to the directrix is a latus rectum, the length of which is $2b^2/a$.

A.2   Conic sections in polar coordinates

For the purposes of dynamics and in particular the orbits of particles under the influence of a central force, it is more convenient to use the polar form of the conic sections rather than the Cartesian form. Thence, the conic section will depend only upon two parameters: the eccentricity e and the semi-latus rectum l. (Recall that the latus rectum is $2l$, (see Section 5.5).)

Consider a conic section with eccentricity e, vertex at $x = 0$ and focus F at $x = -a$ as depicted in Figure A.5. The direction is labelled D with equation

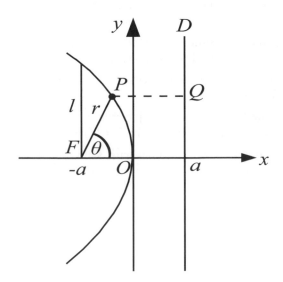

FIGURE A.5
Conic section

$x = a$. Now, for any conic section the eccentricity e is given by

$$e = \frac{|FP|}{|PQ|} = \frac{r}{2a - r\cos\theta}.$$

Rearranging gives

$$r = \frac{2ae}{1 + e\cos\theta}.$$

At $\theta = \pi/2, r = l$; hence, $l = 2ae$ and

$$r = \frac{l}{1 + e\cos\theta},$$

which is the polar form of the equation of a conic section as obtained in Section 5.5 as equation (5.36). The conic is then an ellipse if $0 \le e < 1$, a parabola if $e = 1$ or a hyperbola if $e > 1$.

A.3   Exercises

1. Sketch and identify the conic section with polar equations

   $$r = \frac{4}{1 + 3\cos\theta}$$

   and determine the equation of the directrix. What is the value of the eccentricity?

2. Derive the polar equation of a conic section whose focus is at the origin. The directrix is $y = -2a$ and it has eccentricity e.

3. Consider a parabola. Show that the semi-latus rectum is twice the distance from the vertex to the focus.

4. Consider an ellipse. Show that the semi-major axis a, semi-minor axis b and the semi-latus rectum are related by $l = b^2/a$.

5. Repeat the calculation in $A4$ for a hyperbola with semi-transverse axis a and semi-conjugate axis b.

Solutions

Chapter 1

1. $\dot{x}(t) = \frac{v_0}{1 - v_0 ct}$; $x(t) = (x_0 c - \ln|1 - v_0 ct|)/c$.

2. $x(t) = (2x_0 - v_0)e^t - (x_0 - v_0)e^{2t}$;
   $\dot{x}(t) = (2x_0 - v_0)e^t - 2(x_0 - v_0)e^{2t}$.

3. Max height $= g(1 + \sqrt{3})$; time $= g$; velocity $= -g^2\sqrt{3}$.

4. 100 m.

5. $\frac{192}{v+6}$ seconds; $\frac{60}{v}$ seconds; $v = 10$; $\frac{1}{3}$ ms^{-2}; $\frac{5}{3}$ ms^{-2}.

6. 166.8 kmh^{-1}; on a bearing approx. N20°W.

7. 14 ms^{-1}; $\arctan \frac{11}{5\sqrt{3}}$.

Chapter 2

1. $x(t) = -\frac{1}{m\omega^2}\sin \omega t + C_1 t + C_2$, where C_1 and C_2 are constants.

2. $E = mx^2/2 + kx^3/3$; $x = (3mu^2)^{1/3}$.

3. $x = -a$;

   (i) $-\sqrt{\frac{k}{ma}} < u < \sqrt{\frac{k}{ma}}$;

   (ii) $u \leq -\sqrt{\frac{k}{ma}}$ and $\sqrt{\frac{k}{ma}} \leq u < \sqrt{\frac{2k}{ma}}$;

   (iii) $u > \sqrt{\frac{2k}{mu}}$.

4. The positive potential energy term a/x^{12} describes a repulsive force, while the negative potential energy term $-b/x^6$ describes an attractive force.

   $x = (2a/b)^{1/6}$.

5. Acceleration is $g/5$ ms^{-2}; tension is $48g/5$ N and normal reaction is $36g/5$ N.

6. $1/3$ m.

7. Acceleration is $(6 - 3\sqrt{2})g/10$; tension is $(12 + 9\sqrt{2})g/5$ and force is $9(2\sqrt{3} + 1)g/10$.

Chapter 3

1. -

2. $\frac{1}{2\pi}\sqrt{\frac{g}{y_1 - y_0}}$.

3. $\frac{2ma^3}{k}$; $\frac{m}{18}\sqrt{\frac{2a^7}{b^5}}$.

4. $x(t) = e^{-\frac{bt}{2m}}\left(u_0 \cos\left(\frac{k}{m} - \frac{b^2}{4m^2}\right)^{1/2}t + \frac{2mv_0 - bu_0}{\sqrt{4mk - b^2}}\sin\left(\frac{k}{m} - \frac{b^2}{4m^2}\right)^{1/2}t\right)$;

   amplitude, $A = \left(u_0^2 + \frac{2mv_0 - bu_0)^2}{4mk - b^2}\right)^{1/2}$;

   phase shift, $\phi = \tan^{-1}\left(\frac{2mv_0 - bu_0}{u_0\sqrt{4mk - b^2}}\right)$;

   $A \to \infty$ as $b \to \sqrt{4mk}$.

5. a; π/ω.

6. $x_0\gamma$; $2\pi x_0\sqrt{\frac{m\gamma}{2V_0}}$.

7. $t = \pi$; s_1: $\dot{x}_1(\pi) = +1$ for 10 kg mass; $\dot{x}_2(\pi) = -2$ for 20 kg mass.
   $\therefore t = 2n\pi$ or $t = (2n + 1)\pi/3$.
   At $t = 2n\pi$ both are going up.
   At $t = (2n + 1)\pi/3$ 10 kg mass is going up, 20 kg mass is going down.

8. -

9. $m\ddot{x} + 2am\omega\dot{x} + m\omega^2 x = 0 \Rightarrow \ddot{x} + 2a\omega\dot{x} + \omega^2 x = 0$;
   $\therefore \lambda = -a\omega \pm i\sqrt{1 - a^2} \Rightarrow x(t) = e^{-a\omega t}\frac{u}{\sqrt{1-a^2}}\sin\sqrt{1 - a^2}t$.

10. $x(t) = A_0\cos(\omega t - \phi)$, where $A_0 = \frac{4}{\sqrt{(4-\omega^2)^2 + \omega^2}}$; $\phi = \tan^{-1}\left(\frac{\omega}{4-\omega^2}\right)$
    $A_{max} = 8\sqrt{2/15}$; $\omega = \sqrt{7/2}$.

11. $q(t) = \frac{q_0}{15}e^{-5t}(15\cos 15t + \sin 15t)$.

12. V_0/Z
    N.B. $Z = (x^2 + R^2)^{1/2}$
    $\therefore A_{I(t)} = \sqrt{\frac{V_0^2 R^2}{Z^4} + \frac{V_0^2 X^2}{Z^4}} = \frac{V_0}{Z^2}\sqrt{R^2 + X^2} = \frac{V_0}{Z}$.

Chapter 4

1. $\tan\theta_1 = 2\tan\theta_2$; $u = 44.7$ ms^{-1}; time $= 1.6$ s.

2. $(x, y) = A(ut\cos\theta, ut\sin\theta - 5t^2)$ and $B(ut\sin\theta, ut\cos\theta - 5t^2)$; $3\pi/4$;
   $A\left(\frac{u^2}{20}\sin 2\theta, \frac{u^2}{40}(1 - \cos 2\theta)\right)$; $\pi/8$.

3. $\frac{u}{5}\left(\frac{\sin\theta_2}{\cos\theta_1}\right)$; $\frac{u^2}{5}\left(\frac{\sin(2\theta_2 + \theta_1) - \sin\theta_1}{1 + \cos 2\theta_1}\right)$.

4. -

5. -

6. -

7. -

8. 16.

9. $y^2\sin x + xz^3 - 4y + 2z +$ constant; $4\pi + 15$.

10. $x = \frac{E}{\omega B}(\omega t - \sin\omega t)$; $y = \frac{E}{\omega B}(1 - \cos\omega t)$; $z = 0$.

Chapter 5

1. 702 km.

2. 10323 seconds.

3. 21890 km; 6480 km.

4. -

5. -

6. -

7. $\theta = \sin^{-1}(2/3)$; $\omega = (2g/3b)^{1/2}$.

8. $E = \frac{mv_0^2}{2} - \frac{k}{r_0}$.

9. The orbit is unstable because the equation has exponential solutions.

10. $\ddot{r} - \frac{L^2}{m^2 r^3} = -\frac{\beta}{r^3}$; $\Omega^2 \equiv \frac{m^2\beta - L^2}{L^2}$. Circular orbits occur for $\Omega^2 = 0$. Solutions are $u = \frac{1}{r} = c_1 + c_2\theta$, where c_1 and c_2 are constants of integration. Circular orbits arise only when $c_2 = 0$. Any loss of energy would reduce the angular momentum, which would mean that $\Omega^2 \neq 0$. Hence, the orbit is not stable.

11. $r = \frac{b}{2\sin(\theta/2)}$; speed $= 2W$.

Chapter 6

1. $\mathbf{R} = (\frac{1}{3}, \frac{1}{2}, \frac{2}{3})$.
2. -
3. $-\frac{1}{7}(1,1,1)$; $(-1,-1,-1)$; $(0,0,-2)$.
4. 11.
5. -
6. $m_1 = m_2/3$.
7. Velocities of B_1, B_2 and B_3 are $-v/3$, 0 and $2v/3$, respectively; velocities of B_1 and B_3 are $-v/3$ and $2v/3$, respectively. Comment: Provided the mass of B_3 is greater than that of B_1 and B_2 the outcome of the collisions would make the inclusion of B_2 irrelevant.
8. -
9. Speeds of $P_1, \ldots, P_n$ are $u_n, \ldots, u_1$, respectively.
10. 10 m.
11. -
12. -
13. -
14. $\frac{1}{4}; \frac{3}{4}; \frac{u}{8}; \frac{u}{4}; \frac{3u}{4}$.
15. $v(t) = g(1 - e^{-kt})/k$.
16. Velocity of red car was $m_0 v/(m_0 + kt)$. Velocity of green car was ve^{-kt/m_0}. Green's team was not so clever. Red travels farther because the resistance from the rain is the same for both cars but red has the greater momentum as the rainwater does not drain away from red's car.

Chapter 7

1. (i) $4(m_2 + m_3)l^2$;
   (ii) $(m_2 + m_3)l^2$.
2. $I = \frac{2}{3}ml^2$.
3. (i) $\frac{1}{6}Ml^2(1 - \cos\theta)$;
   (ii) Md^2.
4. (i) $\frac{3}{10}Ml^2$;

(ii) $\frac{3}{20}(4h^2 + l^2)$;

(iii) $\frac{1}{20}(2h^2 + 3l^2)$.

5. (i) $\frac{1}{3}M(b^2 + a^2)$;

(ii) $\frac{4}{3}M(a^2 + c^2)$.

6. -

7. $\tau = 2\pi\sqrt{\frac{l^2 + 3d^2}{3gd}}$.

Chapter 8

1. $(2, -1, 2)$ and $(5, 6, 3)$.

2. $(2t\cos t - 3\sin t, 3\cos t + 2t\sin t, 1 - t)$ and $(-\sin t, \cos t, 0)$.

3. -

4. 340 N and 170 N.

5. -

6. Time: $\frac{\cosh^{-1} 2}{\omega}$; speed: $l\omega$.

7. -

8. -

9. South, a distance $\frac{4}{3g^2}\omega v^3 \sin^2\theta \cos\theta \sin\lambda$.

Chapter 9

1. 1.4×10^6 ms^{-1}.

2. 25 years

3. 1.3×10^{-7} s; 5.3 m.

4. -

5. 1 s; $\frac{\gamma}{2} = \frac{1}{\sqrt{3}}$.

6. $T_1 = T_2'\sqrt{\frac{1-\beta}{1+\beta}}$.

Chapter 10

Chapter 11

Appendix

1. Hyperbola; $x = \frac{4}{3}$; $e = 3$.
2. $r = \frac{2ae}{1 - e \sin \theta}$.

Index